Mathematical Reasoning and Heuristics

Mathematical Reasoning and Heuristics

edited by

Carlo Cellucci
and
Donald Gillies

© Individual author and King's College 2005. All rights reserved.

ISBN 1-904987-07-9
King's College Publications
Scientific Director: Dov Gabbay
Managing Director: Jane Spurr
Department of Computer Science
Strand, London WC2R 2LS, UK
kcp@dcs.kcl.ac.uk

http://www.dcs.kcl.ac.uk/kcl-publications/

Cover design by Richard Fraser, www.avalonarts.co.uk
Printed by Lightning Source, Milton Keynes, UK

All rights reserved. No part of this publication may be reproduced, stored in a retrieval system or transmitted, in any form, or by any means, electronic, mechanical, photocopying, recording or otherwise, without prior permission, in writing, from the publisher.

CONTENTS

Introduction **Carlo Cellucci and Donald Gillies**	vii
List of Contributors	xxi
Constructive Ambiguity in Mathematical Reasoning **Emily Grosholz**	1
Similarities and Differences Between the Development of Geometry and of Algebra **Ladislav Kvasz**	25
Reflections on the Proliferous Growth of Mathematical Concepts and Tools: Some Case Histories from Mathematicians' Workshops **Yehuda Rav**	49
Categorification as a Heuristic Device **David Corfield**	71
Heuristics and Mathematical Discovery: the Case of Bayesian Networks **Donald Gillies**	87
Do We Really Need Axioms in Mathematics? **Gianluigi Oliveri**	119
Mathematical Discourse vs. Mathematical Intuition **Carlo Cellucci**	137
Mathematical Explanation **Mary Leng**	167

Can a Proof Compel Us?
Cesare Cozzo

Introduction
CARLO CELLUCCI AND DONALD GILLIES

This book is based on the papers presented at the conference 'Mathematical Reasoning, Heuristics and the Development of Mathematics', which took place at the University of Rome 'La Sapienza', Villa Mirafiori, on 10-11 September 2004. The announcement of the conference said:

'The conference is intended to cover both general questions in the philosophy of mathematics and special questions concerning the nature of mathematical reasoning, the role of heuristics in mathematical reasoning, and the growth of mathematical knowledge.'

The speakers at the conference were David Corfield, Cesare Cozzo, Brian Davies, Michael Dummett, Donald Gillies, Emily Grosholz, Ladislav Kvasz, Mary Leng, Gianluigi Oliveri, Dag Prawitz and Yehuda Rav. Professor Davies's talk, entitled 'A Defence of Mathematical Pluralism', had already been scheduled for publication in *Philosophia Mathematica*, where it will appear. Pressing engagements prevented Professor Sir Michael Dummett and Professor Dag Prawitz from contributing to the present volume. Papers by the other eight speakers are published here, in some cases with significant changes. Two of these papers (those by Gillies and Oliveri) discussed some views of one of the editors (Carlo Cellucci) which have so far been expounded in detail only in Italian. The other editor (Donald Gillies) therefore persuaded Carlo Cellucci to include a paper summarizing his views. The conference and the publication of this volume were financed by a Cofin 2002 Grant of Miur, the Italian Ministry of University and Research, and the editors would like, on behalf of all the participants, to express their thanks for this support.

The order of the papers in this volume is different from the order in which they were given at the conference since an attempt has been made to rearrange them along thematic lines. The first two papers by Grosholz and Kvasz are mainly concerned with the analysis of pre-twentieth century mathematics (Galileo, Desargues, Cardano, Klein and Galois). Rav's paper is also historical but, as it starts with a consideration of Pythagoras' theorem and ends with Einstein, it may be said to bridge the gap between earlier mathematics and the twentieth century. The next two papers by Corfield and Gillies take us well into the twentieth century, and up to the present. In his recent (2003) book: *Towards a Philosophy of Real Mathematics*, Corfield has urged philosophers of mathematics to devote more attention to the non-foundational mathematics of the last seventy years, and in his paper he obeys his own maxim by analysing higher-dimensional algebra and category theory. Gillies also adopts Corfield's programme, but his example of recent mathematics is quite different. His paper focuses on the discovery of Bayesian networks which took place in the 1980s, and tries to analyse the heuristics which were involved in this advance. The next two papers deal with the question of the value of the axiomatic method for mathematics. Oliveri defends the axiomatic method, while Cellucci criticizes it, and proposes an alternative — the analytic method. The papers so far have been somewhat historical in orientation — provided that we include contemporary developments within history. However the last two papers have a more logico-linguistic approach. Leng criticizes a new version of the Quine-Putnam indispensability argument for the existence of mathematical objects, while Cozzo takes up the question of how exactly proofs compel us to accept their conclusions. He begins by discussing Wittgenstein's and Dummett's views on this question, but goes on to modify these to produce some views of his own. It is interesting to note that Cozzo is rather more historically minded than either Wittgenstein or Dummett since he discusses Plato's example of the slave boy from the Meno, and also discusses Lakatos's views on proof. We will next summarise briefly some of the main ideas in the papers. This will enable us to see if we can detect any general themes in the volume, and to discuss what these themes tell us about the state of philosophy of mathematics at the beginning of the twenty first century.

Let us begin therefore with Grosholz's paper. It is a common enough view that the terms used by mathematicians should be clear and unambiguous, but Grosholz argues to the contrary that ambiguity can play a very constructive part in the development of mathematics. Her paper focuses mainly on an analysis of Galileo's treatment of free fall in the *Two New Sciences*, and she writes (p. 10):

> Galileo is now using at least four modes of representation to express his argument: proportions, geometrical figures, numbers,

and natural language. He also employs a systematic ambiguity to carry his argument further. ... *There is only one set of diagrams, but the set must be read in two ways.* Reading the intervals as finite allows both for the application of Euclidean results, and for the pertinence of the arithmetical pattern just noted. Reading the intervals as infinitesimal allows for the analysis of accelerated motion. The accompanying text in natural language guides and exploits this double meaning.

Grosholz relates this to the views of Carnap, by saying (p. 14):

> The reductionist program set out in Rudolf Carnap's *The Logical Structure of the World* can only conclude that Galileo's text should be re-written.

She argues that this sort of problem has led the heirs of Carnap and Hempel to adopt a more semantic approach and the study of models. She herself, however, proposes to go further and consider pragmatics as well as semantics since (p. 21):

> Focusing on the pragmatic dimension of mathematical language allows us to see the philosophical interest of useful ambiguity in mathematics.

Grosholz shows in conclusion that the use of ambiguity in mathematics is not a thing of the past by giving a striking example from twentieth century mathematics. Ambiguity is involved in Gödel's incompleteness theorems, since (p. 23):

> In Gödel's proof, numbers must stand iconically for themselves in order to allow the application of number theoretic results, and symbolically for well formed formulas (which they represent only by convention) ...

Kvasz's paper is concerned with a comparison between the development of geometry and algebra. However he begins by considering the criticisms of Kant's philosophy of mathematics given by Russell and Carnap, and the recent defence of Kant against these criticisms by Hintikka and Friedman, who have shown that both criticisms make use of formal logic, so that before Frege their arguments were not applicable. Kvasz adds to this debate by suggesting that the discovery of non-Euclidean geometry changed the semantics of mathematical theories by introducing the notion of model. This was what enabled Gauss to make the distinction between pure and applied geometry fifty years before Frege. This example shows, so Kvasz argues, the need for a programme of reconstruction of the semantic development of mathematics. He points out that (p. 32): 'In contemporary mathematics the semantic structure of a mathematical theory is described in the framework of set theory.' However this is not adequate to deal with the semantics of mathematical theories in the pre-Cantorian era before set theory had been thought of. In those historical times, intuitive representations played an important role in the semantics of mathematics, and Kvasz proposes to deal with these using some ideas from Wittgenstein's picture theory of meaning in the *Tractatus* — particularly the idea of pictorial form.

It is worth remarking that Kvasz's historical reconstruction of the semantic development of mathematics cannot be carried out using the methodology of much of contemporary analytic philosophy. Those who apply this methodology characteristically translate everything into Fregean logic and Cantorian set theory, thereby preventing themselves from investigating pre-Cantorian semantic changes. Kvasz's approach also has links with that of Grosholz. Both of them show that to deal with mathematics one needs richer conceptions of meaning than those provided by Tarskian semantics.

Kvasz now proceeds to examine some similarities between the development of geometry and of algebra. His argument is that various stages share the same pictorial form. Thus the pictorial form of Desargues' geometry is called the projective form, and Kvasz argues that the same form is to be found in Cardano's algebra. To solve the cubic equation, Cardano introduced the substitution (1) $x = \sqrt[3]{u} - \sqrt[3]{v}$. Kvasz comments (p. 38):

> The substitution (1) is a great innovation, because it introduces a new representation for the unknown, and so the formula (1) itself can be seen as analogous to Dürer's drawing [see p. 35 — Eds.], as *a representation of a representation*. It represents the same thing, namely the unknown, twice. First it represents the unknown using the letter x (which can be seen as an analogy to the jug in Dürer's drawing) and then as $\sqrt[3]{u} - \sqrt[3]{v}$ (which is an analogy to the picture of the jug in the drawing). Further, there is the sign =, which represents the relation between these two expressions. In Dürer's drawing the eye of the painter (becom-

> ing the centre of projection in Desargues) was the point, which founded the sameness of the jug and its image in the representation. Therefore the sign = is an analogy of the *point of view* in algebra. ... The third aspect, which underlines the analogy between Cardano and Desargues, is the discussion of the *casus irreducibilis*, which finally led to the introduction of the complex numbers. Complex numbers are, in our view, *ideal objects*, just as were the infinitely remote points of projective geometry.

This shows the remarkable detail of the analogy which Kvasz develops between geometry and algebra in this case, and he goes on to give a similarly detailed analysis of an analogy between Klein's Erlanger program and Galois' theory which, he argues, share what he calls the 'integrative form'. Kvasz does not, however, argue that there is perfect correspondence between the semantic development of geometry and of algebra, and, in the closing pages of his paper discusses some of the differences between the two cases.

Rav deals with the growth of mathematics using a botanical analogy (pp. 50-51):

> Mathematics grows like a tree, through (a) the dynamic interaction of internal ontogenetic factors — read, internal theoretical developments, (b) nutriments supplied by the environment — read, demands and catalyzing stimulation coming from the various sciences; (c) being planted in the right soil — read, culture.

He illustrates this by a series of interconnected case histories from the history of mathematics, beginning with the Pythagorean theorem where Euclid's strictly deductive argument is contrasted with pictorial demonstrations based on folding to be found in Indian and Chinese mathematics. He then continues with the Pythagorean relation in Cartesian geometry, in calculus, in Gauss's differential geometry, in Riemann's non- Euclidean geometry, and in tensor analysis. In this way Rav shows how the theory of tensors was developed to the point where it could be used by Einstein and Grossman in the General Theory of Relativity.

Rav's paper has taken us into the twentieth century and Corfield's brings us up to the present. He writes (p. 72):

> In recent decades, with the proliferation of mathematics, the issue that has come to the fore has been how best to organise

the field.

Corfield's answer is 'higher-dimensional algebra', and he says of this (p. 74):

> In higher-dimensional algebra, also known as higher-dimensional category theory, you encounter a ladder which you're irresistibly drawn to ascend.

It this way one arrives at 2-categories, 3-categories, ... and so of course (p. 76):

> Mathematicians hate to stop a good thing when it's rolling, so are aiming to extend this process infinitely far to omega-categories.

Corfield is undoubtedly right that there is in contemporary mathematics a seemingly irresistible tendency to move to ever greater abstractions. However some might see this as a bit problematic. Have some of these very abstract mathematical theories lost all contact with material reality? Will they ever be applied, or are they just a modern form of metaphysics? The issue is a tricky one since undoubtedly the attempt to limit mathematics to problems of practical use would have been counter-productive even from the point of view of practice. Thus Apollonius of Perga's study of ellipses in ancient times proved very useful to Kepler in the 17th century — though Apollonius could hardly have foreseen this. Or, to take a more recent example, the mathematical logic of Frege and Russell was developed for purely philosophical reasons, as part of their logicist programme to show that mathematics could be reduced to logic. However their mathematical logic proved very useful in computer science, something which Frege and Russell certainly did not expect since the computer had not been invented when they did their work on logic. Examples like this suggest that mathematics which is developed for purely theoretical reasons may turn out in the fullness of time to have useful practical applications. Yet does this hold for all pure mathematics? Might some parts of contemporary mathematics be just too abstract to find any future application? Corfield anticipates this criticism because he writes (p. 76):

> ... computer scientists from the French nuclear industry are using omega-categories to analyse potential deadlocks in multi-processor computations.

Gillies begins his paper by acknowledging his debt to recent books by Cellucci and Corfield. From Cellucci he takes the idea of studying the heuristics of mathematical discovery, and from Corfield the idea that philosophers of mathematics should consider non-foundational mathematics of the last seventy or so years. While Corfield considers the abstractions of omega-categories, however, Gillies deliberately chooses a rather more practical mathematical innovation — the discovery of Bayesian networks which took place in the 1980s as the result of some problems which had emerged in the development of artificial intelligence. Gillies gives a brief historical sketch of this discovery, and then tries to analyse the heuristic principles which were involved. He claims to find three heuristics, namely (a) the use of philosophical ideas, (b) new practical problems, and (c) domain interaction. He argues that these heuristics, though by no means universal in the development of mathematics, are to be found in many other cases of mathematical discovery. His paper concludes by asking whether the heuristics of mathematical discovery can be said to constitute a logic of mathematical discovery. This is not an easy question to answer because of vagueness concerning the limits of what constitutes logic.

The next two papers by Oliveri and Cellucci are concerned with the axiomatic method in mathematics. Oliveri's paper is a defence of the use of this method. He stresses the very widespread application of the axiomatic method in mathematics which has occurred since the time of Hilbert. However, objections to it can be raised from the limitative results of Gödel and Tarski. Oliveri defends the axiomatic method against the claim that it has been damaged by these results, and also against the more general objections of Brouwer and Cellucci. Contrary to Cellucci, Oliveri argues that the axiomatic method can be heuristically fruitful, and he cites (pp. 128–9) the example of investigations into the continuum hypothesis. These began with Cantor's initial attempts to prove the continuum hypothesis. However a significant advance was only made after set theory had been axiomatised by Zermelo, Skolem, and Fraenkel. This axiomatisation enabled Cohen to prove the independence of the continuum hypothesis from the other axioms of set theory, shedding a light on the question which would not have been possible had the axiomatic method not been applied to set theory.

Cellucci in his paper puts forward a series of objections to the axiomatic method and criticizes the attempts which have been made to overcome these difficulties. The first problem he considers is that the axiomatic approach does not provide any specific content for mathematics, or, as he puts it (p. 137):

> One of the most uninformative statements one could possibly make about mathematics is that the axiomatic method expresses the real nature of mathematics, i.e., that mathematics consists in the deduction of conclusions from given axioms. For the same could be said about several other subjects, for example, about theology.

Cellucci criticizes in detail Hintikka's recent attempt to overcome this objection using second order logic.

Cellucci's next problem is (p. 140): 'How are axioms justified?' The traditional answer is: 'by intuition', to which Cellucci replies (p. 140): 'But all known justifications of axioms in terms of intuition are inadequate.' Cellucci elaborates this by criticizing in detail the uses of intuition made by Gödel and Hilbert. Cellucci sees little more hope of justifying axioms on the grounds that true consequences follow from them. This brings us to the problem of how axioms are discovered. Cellucci thinks that this can be solved by the analytic method which is indeed his alternative to the axiomatic method. The following is a brief description which he gives of the analytic method (p. 149):

> To solve a mathematical problem, we formulate a hypothesis that is a condition sufficient for its solution. The hypothesis is obtained from the problem, and possibly other data, by non-deductive inferences — inductive, analogical, etc. However, the hypothesis must not only be a condition sufficient for the solution of the problem but must also be plausible. That is, it must be compatible with the existing knowledge, ...

It is also important to note that (p. 152): 'The analytic method is both a method of discovery and a method of justification.'

Once a hypothesis has been proposed as a solution to a problem, it becomes itself a problem, and so, continuing to use the analytic method, a new hypothesis should be proposed to explain the old hypothesis and so on indefinitely. This indefinitely continuing procedure is what distinguishes the analytic method from the axiomatic method, for, as Cellucci says (p. 154):

> Actually, the axiomatic method is what results from the analytic method if, at a certain stage, the process of passing from one hypothesis to another is stopped definitively, the hypothesis reached at that step being considered no longer as a problem to be solved but as an absolutely unproblematic starting point. Therefore the axiomatic method is an unjustified truncation of the analytic method which removes the most important part of it.

One objection which might be raised to Cellucci's analytic method is that it relies on non-deductive inferences and that these are more dubious and less reliable than deductive inferences. To meet this point, Cellucci concludes his paper by arguing that demonstrative reasoning is no better than non-demonstrative reasoning, since (pp. 162):

> ...it is generally impossible to know whether the premises of demonstrative argument are true...but also...it is generally impossible to verify the correctness of demonstrative arguments, even when they are obtained by formal inference rules. For some demonstrative arguments are so long and complex that one can never be sure that they contain no mistakes.

Leng's paper moves onto terrain which is perhaps more familiar to philosophers of mathematics in the English-speaking world. She begins by saying (p. 167):

> In this paper, I would like to ... consider ... some philosophical issues that arise when we use mathematical reasoning in the natural sciences to explain empirical phenomena.

Leng is concerned with a new form of the Quine–Putnam indispensability argument which has been put forward recently by Baker and Colyvan and which has the advantage that it avoids an appeal to confirmational holism. This form of the argument rests on a consideration of some examples of the use of mathematics to explain empirical phenomena. If these explanations are satisfactory, as they appear to be, then we should regard the explanans as true, and this, in the usual Quine–Putnam fashion, commits us to the existence of mathematical objects. Leng is a supporter of anti-realism regarding mathematical objects, however, and she counters this argument by saying that the mathematics appears only in models and the successful use of models does not commit us to the existence of the entities of the models. Regarding models, Leng is undoubtedly correct. Consider for example a model of a complicated molecule made of wooden balls connected by springs. Suppose the wooden balls representing oxygen are painted red. No one would maintain that the success of such a model shows that oxygen is really red in colour. On the other hand, if we use a model which involves mathematics are we to say that the mathematics is one of the fictional elements in the model like the red paint on the oxygen balls? It is not clear that this is the correct response. After all, in the example of the model of the complicated molecule, it is precisely the geometrical relations between the wooden balls which do correspond to reality, since, if the model is a good one, they accurately represent the geometrical relations holding between the atoms of the actual molecule.

The final paper of the collection by Cozzo deals with the important problem of analysing the compulsion which good mathematical proofs appear to have. Of course this compulsion could be dismissed as a psychological illusion. As Cozzo says (p. 198):

> If the objectivity of good inference is abandoned, the thesis of compulsion is false. We cannot anymore distinguish between an inference which is compelling for S and an inference which S is only inclined to accept. The whole idea of compulsion of proof would reduce to the fact that we sometimes *feel* compelled or *believe* ourselves to be compelled.

Cozzo, however, does not abandon the objectivity of good inference and so has to give an account of the compulsion of proofs. He first considers what he calls 'the Wittgensteinian critique' — though he adds that it may partly be due to Dummett. Essentially (p. 198): 'The Wittgensteinian Critique does not deny compulsion, but reduces it to a social compulsion.' The trouble with this approach is that it may bind a particular individual but does not bind the community as a whole. As Cozzo says (p. 200):

> If good inference is relative to a community, and being a good inference is the same as being accepted by the community, no genuine normativity holds for the community: the community can never be wrong.

Cozzo regards it as quite counterintuitive to say that a community can never discover an objective mistake so that we actually need (p. 201): 'a *community-independent objectivity of good inference.'*

But what kind of constraint can limit the community's freedom? Cozzo considers Dummett's answer that the constraint is provided by the meanings of the involved expressions. However, this runs into difficulties because of the 'plasticity of meaning'. Moreover Cozzo argues with a reference to Lakatos' *Proofs and Refutations* that (p. 209):

> In the course of mathematical rational controversy accepted meanings (i.e. concepts) can be criticized and new meanings (i.e. concepts) can be put forward.

This leads Cozzo to his conclusion which in some ways sounds like a synthesis of Wittgenstein and Popper. It runs (p. 211):

> Inferential steps can be forced on us, but they are not forced by a necessity external to us: they are forced by our common critical activity.

Let us now turn to analysing some of the main themes to be found in this volume and to considering what they show regarding the present state of philosophy of mathematics. Philosophy of mathematics was dominated for most of the twentieth century by the three classic schools which flourished in the period 1900 to 1930, namely logicism, formalism and intuitionism. Yet arguably these three schools had all been shown to be inadequate by the early 1940s if not before. Gödel's incompleteness theorems do appear to refute logicism and formalism fairly conclusively — despite claims to the contrary. Of course intuitionism was never refuted in quite that way — although some have maintained that it too is affected by Gödel's incompleteness theorems. Yet Brouwer's original hope was that intuitionistic mathematics would replace mainstream mathematics, and this did not happen. Neither intuitionism nor any other variety of constructive mathematics has replaced standard mathematics. Despite this, one continues to the present day to find philosophers of mathematics working on varieties of neo-logicism, neo-formalism and neo- constructivism. Are these not the ghosts of departed research programmes?

In the light of all this it is interesting to note that the present collection does not contain a single paper which is concerned with any of the problems which were raised in the three classic schools of philosophy of mathematics. Instead we find discussion of a new range of problems. Many of these new problems can be traced back to the work of Lakatos, Quine, and Wittgenstein which appeared after the Second World War. However the formulations of these problems and the way they are discussed show very considerable developments relative to the work of these three important thinkers.

The three classic schools all took as central the problem of justifying the mathematics of their day (c. 1900 — c. 1930). Of course this formulation needs to be modified in the case of intuitionism which regarded contemporary mathematics as unjustifiable and therefore sought to replace it by a mathematics which could be justified. Still the issue of justification was as central for intuitionism as for the other two schools. Of course the method of justification varied from school to school. Logicists proposed justifying mathematics by reducing it to logic, formalists proposed justifying mathematics by reformulating it as a series of formal systems each of which would be proved consistent at the meta-level using only incontrovertible finitistic methods, while intuitionists wanted to justify mathematics in terms of the constructions of the creative mathematician. By contrast none of the papers in the present volume, with the possible, but partial, exception of Cozzo's, even raise the question of justification in the traditional sense.

Another striking difference is that the three classic schools identified mathematics with the mathematics of their own period say c. 1900 to c. 1930. By contrast most of the contributors to the present volume take for granted that mathematics has developed historically and that there are significant differences between the mathematics of different historical periods. This historical approach was obviously stimulated by the work of Lakatos.

Once mathematics is viewed historically new questions arise concerning how the growth of mathematics occurred. One aspect of this growth is changes in the meaning of the terms used by mathematicians. This semantic aspect of mathematical development is studied by Grosholz and Kvasz. Both of these writers show the inadequacy of standard Tarskian semantics for dealing with this problem. Grosholz, through a study of Galileo, shows the creative power of ambiguity in the development of mathematics; while Kvasz constructs a framework for studying the semantic aspects of mathematics in the pre-set-theoretic period. His concept of 'pictorial form' was originally developed for geometry but in his paper in this volume he shows that it can be applied to algebra as well. Rav's account of the development of geometrical ideas from Pythagoras' theorem does indeed demonstrate a proliferous growth of mathematical concepts and tools.

One way of escape from the mathematics of the period c. 1900 to c. 1930 is to move backwards in time, but another is to move forwards and consider some of the mathematics which has developed since 1930. This is the approach taken by Corfield and Gillies. Corfield discusses the contemporary theory of higher-dimensional categories, while Gillies considers the theory of Bayesian networks which was discovered in the 1980s.

The title of Oliveri's paper: 'Do we really need axioms in mathematics?' formulates a question which was hardly considered in the three classic schools. It is true that the question might have been partially raised by the dissident Brouwer, but other intuitionists such as Heyting had no qualms about the axiomatic method. As for logicists and formalists, they regarded the axiomatic method as so central that they would never have raised any questions as to its desirability. Oliveri, however, writing today, feels that the axiomatic method is in need of a defence. This is not surprising since it has come under attack from Cellucci whose objections to the axiomatic method and suggestions for an alternative 'analytic method' are set out in his paper.

Another feature of the three classic schools is that their proponents rarely considered the question of how mathematics is applied. Their attention was mainly fixed on pure mathematics. Issues to do with the application of mathematics were, however, introduced by Quine and Wittgenstein, and we find discussions of applied mathematics throughout the present volume. Grosholz's example of Galileo's mathematical treatment of free fall is obviously an example of mathematics applied to empirical material. Interestingly the two authors who deal with post-1930 mathematics, both mention applications to computer science. This is indeed a significant development which has occurred in the last few decades. Historically the main area of application of mathematics was physics, but now mathematics is applied also to computer science. Applications in this new field have a character which is quite significantly different from applications in physics. The same could be said about applications of mathematics to biology, epidemiology, linguistics, etc.

Leng considers another philosophical use of applied mathematics, namely the argument, which stems from Quine, that successful applications of mathematics in physics show that mathematical objects, such as numbers, are just as real as physical objects, such as electrons. Of course Leng, being an anti-realist, does not accept this line of argument.

Cozzo's analysis of the nature of the compulsion produced by mathematical proofs might be considered as to do with an issue of justification, but this question of justification is hardly the same as those raised by the three classic schools. The centre of Cozzo's investigation is to do with the sociological approach to mathematics introduced by the later Wittgenstein.

So we see that the present volume deals with a series of questions quite different from those which occupied the minds of the proponents of the three classic schools of philosophy of mathematics — logicism, formalism, and intuitionism. These questions are no longer to do with justification in the traditional sense, but with a variety of other topics. Some are concerned with discovery and the growth of mathematics. How does the semantics of mathematics change as the subject develops? What heuristics are involved in mathematical discovery, and do such heuristics constitute a logic of mathematical discovery? What new problems have been introduced by the development of mathematics since the 1930s? Other questions are concerned with the applications of mathematics both to physics and to the new field of computer science. Then there is the new question of whether the axiomatic method is really so essential to mathematics as is often supposed, and the question, which goes back to Wittgenstein, of the sense in which mathematical proofs are compelling. Taking these questions together — to which others could be added, such as the question of how mathematics is related to our brain, or neural activity, or cognitive architecture, which has received considerable attention recently — we can see the outline of a new agenda emerging which is likely to carry philosophy of mathematics forward into the twenty first century.

Carlo Cellucci and Donald Gillies
April 2005

List of Contributors

IN ALPHABETICAL ORDER

CARLO CELLUCCI

Carlo Cellucci graduated in philosophy at the University of Milan in 1964, writing a dissertation entitled *Ordinali ricorsivi* [Recursive Ordinals] with Ludovico Geymonat as supervisor. After spending two years in Oxford as a National Research Council and Royal Society fellow, working on intuitionistic logic and mathematics under the guidance of John N. Crossley and Michael Dummett, he was lecturer in Mathematical Logic at the University of Sussex, lecturer in Logic at the University of Siena, professor of Philosophy of Science at the University of Calabria and professor of Philosophy of Science at the University of Siena. From 1979 he has been professor of Logic at the University of Rome 'La Sapienza'. His research has been in mathematical logic, especially proof theory, in the philosophy of mathematics and more recently in epistemology. On proof theory he has published the book *Teoria della dimostrazione* [Proof Theory] (Boringhieri, Turin 1978). On the philosophy of mathematics he has edited two collections: *La filosofia della matematica* [The Philosophy of Mathematics] (Laterza, Bari 1967), and *Il paradiso di Cantor* [Cantor's Paradise] (Bibliopolis, Naples 1979). In the past few years he has published two books, *Le ragioni della logica* [The Reasons of Logic] (Laterza, Bari 1998) and *Filosofia e matematica* [Philosophy and Mathematics] (Laterza, Bari 2003) where he develops an alternative to the prevailing orthodoxy in logic and in the philosophy of mathematics. At present he is writing a book, *Filosofia e conoscenza* [Philosophy and Knowledge], where he plans to develop an alternative to the prevailing orthodoxy in epistemology. He has been an editor of the series *Studies in Proof Theory* (Bibliopolis, Naples) and *Instrumenta Rationis: Sources for the History of Logic in the Modern Age* (Clueb, Bologna). He has also edited the following conference proceedings: (with E. Agazzi) *Logiche moderne* [Modern Logics] (Istituto della Enciclopedia Italiana, Rome 1981); (with G. Sambin) *Temi e prospettive della logica e della filosofia della scienza contemporanee* [Themes and Perspectives in Contemporary Logic and Philosophy of Science] (Clueb, Bologna 1988); (with Maria Concetta Di Maio and Gino Roncaglia), *Logica e filosofia della scienza: problemi e prospettive* [Logic and Philosophy of Science: Problems and Perspectives] (ETS, Pisa 1994); (with Vito Michele Abrusci, Roberto Cordeschi and Vincenzo Fano), *Prospettive della logica e della filosofia della scienza* [Perspectives in Logic and in the Philosophy of Science] (ETS, Pisa 1998).

Department of Philosophical and Epistemological Studies,
Faculty of Philosophy,
University of Rome 'La Sapienza',
Via Carlo Fea 2,
00161 Rome, Italy.
Email: carlo.cellucci@uniroma1.it

DAVID CORFIELD

David Corfield studied mathematics at Cambridge as an undergraduate, taking Part III of the Tripos. After a spell in Paris studying psychoanalysis, he began graduate studies in the philosophy department at King's College London in 1991, and he completed his PhD on practice-oriented approaches to the philosophy of mathematics in 1995 with Professor Donald Gillies as supervisor.

In the past 10 years, he has had a number of teaching and research posts in Leeds, London, Cambridge, Oxford, York and Tübingen. His research has been in the history and philosophy of mathematics and probability theory, and in the philosophy of psychology. With Jon Williamson he edited the collection *Foundations of Bayesianism* (Kluwer, 2001) and has recently published *Towards a Philosophy of Real Mathematics* (Cambridge 2003). In 2005, *Why do we get ill?*, jointly written with Darian Leader, will appear with Penguin Books.

Department of Philosophy,
University of York,
Heslington,
York YO10 5DD, UK.
Email: dc23@york.ac.uk

CESARE COZZO

Cesare Cozzo studied philosophy at the University of Rome "La Sapienza" as an undergraduate. In 1987 he began graduate studies at the University of Florence and in 1992 he completed his first PhD thesis *Teoria del significato e filosofia della logica* (later published by Clueb, Bologna) with Professors Carlo Cellucci, Marisa Dalla Chiara and Paolo Parrini as supervisors. In 1995 Cozzo received a second doctorate in theoretical philosophy at the University of Stockholm with Professor Dag Prawitz as supervisor. The Swedish dissertation, *Meaning and Argument*, Almqvist & Wiksell, Stockholm, 1994, presents a theory of meaning centred on the notion of "immediate argumental role". The immediate argumental role, constitutive of understanding, is a particular aspect of the use of a sentence in arguments, but it is not the whole use in arguments, nor is the whole use in arguments reducible to the immediate argumental role. Unlike current inferential conceptions of meaning, this theory distinguishes between the *understanding* and the *correctness* of a language and thus rejects the idea that there can be analytic truths.

From 1998 to 2001, Cesare Cozzo was research fellow at the University of Rome "La Sapienza", and since 2001 he has been associate professor. His research has been focused on the relations between the philosophy of logic and mathematics and the theory of meaning. He wrote several articles on analytic philosophy, the paradox of knowability, the epistemic conception of truth and holism, which appeared in *Erkenntnis, Rivista di filosofia, Theoria, Topoi* and other journals or collected volumes. At present, he his writing a book on the philosophy of Michael Dummett.

Department of Philosophical and Epistemological Studies,
Faculty of Philosophy,
University of Rome "La Sapienza",
Via Carlo Fea 2,
00161 Roma, Italy.
Email: Cesare.Cozzo@uniroma1.it

DONALD GILLIES

Donald Gillies studied mathematics and philosophy at Cambridge as an undergraduate. In1966 he began graduate studies in Professor Sir Karl Popper's department at the London School of Economics, and he completed his PhD on the foundations of probability in 1970 with Professor Imre Lakatos as supervisor.

From 1968 to 1971, he was a Fellow of King's College, Cambridge, and since 1971 he has been a member of staff of London University, though he has changed colleges starting in the now defunct Chelsea College, moving from there to King's College London, and, in October 2004, to University College London.

His research has been in the history and philosophy of science and mathematics with a particular interest in probability theory. He edited the collection: Revolutions in Mathematics (Oxford, 1992), and his most recent book is Philosophical Theories of Probability (Routledge, 2000)

Department of Science and Technology Studies,
University College London,
Gower Street,
London WC1E 6BT, UK.
Email: donald.gillies@ucl.ac.uk

Emily Grosholz

Emily Grosholz studied mathematics and philosophy at the University of Chicago as an undergraduate. She began graduate studies at Yale University in 1973, and wrote a dissertation on the growth of mathematical knowledge with Angus Macintyre and Stephan Koerner, obtaining her PhD in Philosophy in 1978. Since 1979, she has taught in the Philosophy Department at the Pennsylvania State University, where she is Professor of Philosophy and a Fellow of the Institute for Arts and Humanities. She has been a visiting scholar at the University of Pennsylvania, the University of Toronto, the University of Cambridge, and the University of Paris; and a fellow at the National Humanities Center and Clare Hall, University of Cambridge. She has been awarded fellowships from the Guggenheim Foundation, the NEH, and the ACLS, and a Transatlantic Cooperation Grant from the Humboldt Foundation with Herbert Breger.

Her research has been in the history and philosophy of mathematics. Her books include *Cartesian Method and the Problem of Reduction* (OUP, 1991), *Leibniz's Science of the Rational* with E. Yakira (Studia Leibnitiana, Sanderheft 26, 1998) and *Reduction and Representation*, as well as an edited volume with H. Breger, *The Growth of Mathematical Knowledge* (Kluwer, 2000).

Department of Philosophy,
240 Sparks Building,
The Pennsylvania State University,
University Park,
PA 16802, USA
Email: erg2@psu.edu

LADISLAV KVASZ

Ladislav Kvasz graduated in 1986 in mathematics at the Comenius University in Bratislava. In 1989 he spent five months on a scholarship at Moscow State University with Professor D. Sokoloff, where he worked on asymptotic methods in astrophysics. After the changes of the year 1989 he began graduate studies in philosophy at the Comenius University in Bratislava with Professor Miroslav Marcelli as supervisor. In May 1995 he defended his thesis "Classification of Scientific Revolutions". Since 1986 he has been employed at the Faculty of Mathematics and Physics of Comenius University, Bratislava as a lecturer. He gives courses in history and philosophy of mathematics. In 1993 he won a Herder Scholarship and spent the academic year 1993/94 at the University of Vienna studying the philosophy of the Vienna Circle and of Ludwig Wittgenstein. In 1995 he won a Masaryk Scholarship of the University of London and spent the academic year 1995/96 at King's College London working with professor Donald Gillies. In 1997 he won a Fulbright Scholarship and spent the summer term of the academic year 1998/99 at the University of California at Berkeley, working with Professor Paolo Mancosu. In 2000 he won a Humboldt Scholarship and spent the years 2001 and 2002 at the Technical University in Berlin working with Professor Eberhard Knobloch. His research has been in the history and philosophy of science and mathematics, with particular interest in their cultural background in arts, literature, and theology. He was the co-editor of Appraising Lakatos (Kluwer 2002) and has published a book on the classification of scientific revolutions in Slovak (Gramatika zmeny, Chronos Publishers, Bratislava 1999).

Department of Algebra, Geometry and Mathematics Education,
Comenius University Bratislava,
Mlynska dolina,
84248 Bratislava,
Slovak Republic
Email: kvasz@fmph.uniba.sk

MARY LENG

Mary Leng studied mathematics and philosophy as an undergraduate at Balliol College, Oxford. In 1996 she won a Canadian Rhodes Scholars' Foundation Scholarship, which enabled her to go to Canada to continue her studies as a graduate student in philosophy at the University of Toronto. In Toronto she worked, under the supervision of James R. Brown and Ian Hacking, on a PhD defending anti-realism in the philosophy of mathematics. This work included a case study of mathematical proof, which discussed a research seminar on the classification of C^*-algebras held by George A. Elliott at Toronto's Fields Institute for Research in Mathematical Sciences.

In 2001–2002, she held a postdoctoral position at Toronto, before coming to St John's College, Cambridge in 2002, where she is currently a research fellow in philosophy. She has also held visiting fellowships at the University of California at Irvine (where, together with Mark Colyvan, she was invited to spend a term working with Penelope Maddy on the philosophy of applied mathematics), and at the Peter Wall Institute for Advanced Study, at the University of British Columbia.

Her research while in Cambridge has focused on providing a defence of fictionalism in the philosophy of mathematics. This research is due to appear in 2006 as a book, *Mathematics and Reality*, which is under contract with Oxford University Press. She is also currently editing (with Alexander Paseau and Michael Potter) a collection on *Mathematical Knowledge*, which will also appear in 2006.

St John's College,
Cambridge CB2 1TP, UK.
Email: MCL33@cam.ac.uk

GIANLUIGI OLIVERI

Gianluigi Oliveri studied philosophy at the University of Bari (Italy) as an undergraduate. He received a D.Phil. in philosophy at Oxford with a dissertation on *The principles of analytical philosophy* under the supervision of Professor Sir Michael Dummett. He has held posts at the Universities of Leeds, Keele and Oxford, and is presently at the University of Palermo (Italy). His research has been in the philosophy of mathematics and in the philosophy of language with a particular interest in the realism/anti-realism debate in the philosophy of mathematics and in the impact that naturalism has in the philosophy of language. He edited the collections: *The Philosophy of Michael Dummett* (Kluwer, 1994), *Truth in Mathematics*, (OUP, 1998), and *From the Tractatus to the Tractatus and Other Essays* (Peter Lang, 2001). He is currently writing a book in which he develops a realist philosophy of mathematics.

Università di Palermo,
Dipartimento di Filosofia, Storia e Critica dei Saperi,
90128 Palermo, Italy.
Email: gianluigi.oliveri@unipa.it

Yehuda Rav

Yehuda Rav was born in Vienna, Austria and grew up in Israel, studied mathematics and biophysics at Columbia University, shuttling between New York and Princeton to complete his doctoral dissertation in automata theory under the direction of John von Neumann. From 1951 to 1967 he lived in New York, and since then, has lived in Paris. He held lectureships and professorial appointments at Columbia University, Hofstra University, and as of 1967, at the Mathematics Institute of the University of Paris-Sud (Orsay), until having reached statutory retirement age in 1995. He held visiting professorships at New York University, Charles University in Prague, the Jannus Pannonius University in Pécs (Hungary), and the Johannes Kepler University in Linz, Austria. He was a senior fellow at the Pittsburgh Center for the Philosophy of Science and was affiliated with the Konrad Lorenz Institute for Evolution and Cognitive Studies of the University of Vienna. He is a member of the American Mathematical Society and an Honorary Member of the Austrian Society for Cybernetics Studies. His publications cover the fields of cybernetics, mathematical logic and the philosophy of mathematics. Since his retirement, he has been particularly active in lecturing at seminars and continues writing reviews, with over 200 reviews published to date in Mathematical Reviews.

83, rue de la Santé,
F-75013 Paris,
France.
Email: Yehuda.Rav@wanadoo.fr

Constructive Ambiguity in Mathematical Reasoning

EMILY GROSHOLZ

Argument that employs controlled and highly structured ambiguity can play a central role in mathematical discovery and justification. The exposition of projectile motion given on the Fourth Day, following preliminaries on the Third Day, of Galileo's *Discourses and Mathematical Demonstrations Concerning Two New Sciences*[1] is a good illustration of my central claim; I take it as a case study in the first section of this essay. In the analysis of free fall on the Third Day, the use of proportions is polyvalent, because Galileo asks us to read their terms both as finite and as infinitesimal. When we read them as finite, they allow for the application of Euclidean results and also exhibit patterns among whole numbers; and their configurations stand iconically for geometrical figures. When we read them as infinitesimal, they allow for the elaboration of the beginnings of a dynamical theory of motion, leading to the work of Torricelli and Newton; and their configurations stand symbolically for dynamical, temporal processes. In the exposition of projectile motion, the curve of the semi-parabola, read iconically, stands for a temporal, dynamical process that we "see" whenever a projectile leaves a trail behind it; read symbolically, it stands for an infinite-sided polygon that articulates the rational relations among an infinite array of instances of uniform motion that compose the uniformly accelerated motion of the projectile. And the rationality of that reduction is justified by results involving proportions and the similarity of geometric figures.[2]

---

[1] Galileo Galilei *Dialogues Concerning Two New Sciences*, translated by Henry Crew and Alfonso de Salvio, New York: Dover, 1914/1954.

[2] The philosophical use of the polar terms "icon" and "symbol" is due to C. S. Peirce, who distinguished the former as similar to their objects, and the latter as linked to their objects only by convention. In recent essays, I have made use of the distinction, while insisting on the iconic dimensions of symbols and the symbolic dimensions of icons.

I argue further that Galileo's use of ambiguous modes of representation is typical of reasoning in mathematics, even though the pattern has not been sufficiently noted and studied by philosophers of mathematics. Under the influence of the Vienna School, Anglophone philosophers often write as if the language of mathematics and science is, or ought to be, univocal and transparent; the second section of my essay examines this thesis, which I contest, in the writings of Carnap. The terms of an ideal language, Carnap argues, should refer one-to-one to all and only those things that exist, and its predicates and relations should follow suit. Thus, its locutions should not refer to more than one state of affairs at a time, and should not add anything to the situation: there should be no linguistic "artifacts."

In the third section, I locate my own position in the context of a general tendency in Anglophone philosophy of science and mathematics to move from a syntactic approach to a semantic and indeed pragmatic approach, which studies the use of language in terms of its representational role in an historical context of problem-solving. The 'pragmatic' philosophers find that problem-solving typically requires the juxtaposition of a variety of modes of representation; I emphasize that in such contexts a single mode of representation, used iconically for one purpose and symbolically for another, may be called upon to mean more than one thing. The resultant polysemy generates not confusion but insight. In the fourth section, I argue for taking a pragmatic (as well as syntactic and semantic) view of the logicist program, whose early twentieth century domination of the philosophy of mathematics has still not been sufficiently challenged by philosophers.[3] Then we find that Russell missed the mathematical import of his own work, and note that its greatest mathematical challenge came from Kurt Gödel, who achieved his results by articulating and exploiting the determinate and ineluctable ambiguity of both logical formulae and numbers.

---

[3] Carlo Cellucci, in his recent books, *Filosofia e matematica* (Rome: Laterza, 2002) and *Le Ragioni della Logica* (Rome: Laterza, 1998), has forcefully contested the lingering logicism that skews contemporary philosophy of mathematics, but due to the slowness of the translation of his work into English, its impact is still to be registered. See his essay "The Growth of Mathematical Knowledge: An Open World View," in E. Grosholz and H. Breger, *The Growth of Mathematical Knowledge* (Dordrecht: Kluwer, 2000), pp. 153-176; and "'Introduction' to *Filosofia e matematica*" in R. Hersh (ed.), *Unconventional Essays on the Nature of Mathematics*, New York: Springer-Verlag, forthcoming.

1 Constructive Ambiguity in Galileo's Demonstration of Projectile Motion

Mathematics often requires the combination of different modes of representation in the same argument: equations, diagrams, matrices, tables, proportions, schemata, natural language. Arguments in mathematics do many things. They defend definitions, constitute problems, explain problem solutions, deploy and exhibit methods, and formally or informally present proofs. When modes of representation are combined in mathematical arguments, they may be juxtaposed or superimposed, or carefully segregated to exhibit certain features of the situation. Some arguments — and I claim this is true of Galileo's reasoning examined in this section — may require that one and the same representation be used ambiguously in order for the mathematician to exhibit a novel organization and exploration of things, and for the reader to follow the reasoning.

Galileo's treatment of free fall and projectile motion occurs in the Third Day and Fourth Day of his *Discourses and Mathematical Demonstrations Concerning Two New Sciences* (referred to hereafter as the *Discorsi*). The Third Day of the *Discorsi* is entitled 'Change of Position,' and its first section is 'Uniform Motion.' Galileo defines uniform motion—straight line motion at a constant speed—as 'one in which distances traversed by the moving particle during any equal intervals of time, are themselves equal,' and adds that the equal intervals must be thought of as being arbitrarily chosen; he thus includes the possibility that they may be chosen to be arbitrarily small. The first diagram he offers, Figure 1 which accompanies Theorem I, Proposition I of 'Uniform Motion,' consists of two horizontal lines, the line IK representing time and the line GH representing distance that is re-conceptualized to mean displacement, since we are instructed to suppose that a moving particle is traversing it.[4]

The two lines therefore have a different status, since no particle traverses the time-line. Both lines are, however, measured off in intervals: the left-hand half of line IK is measured in intervals of length DE and the right-hand half in intervals of length EF, while the left-hand half of line GH is measured in intervals of length AB and the right-hand half in intervals of length BC.

Theorem I, Proposition I states, 'If a moving particle, carried uniformly at a constant speed, traverses two distances the time-intervals required are

---
[4]Ibid., pp. 153–156.

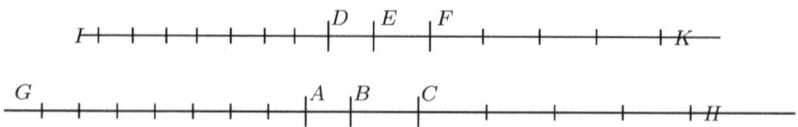

Figure 1.

to each other in the ratio of these distances.'[5] This theorem asserts that a (non-continuous, that is, without a shared middle term between the two ratios) proportionality $AB : BC :: DE : EF$ holds between any two displacement intervals and any two corresponding time-intervals in uniform motion. Galileo has designed the diagrams and the reasoning to allow for a direct application of the Euclidean/Eudoxian axiom, which states that proportions between non-continuous ratios, $A : B :: C : D$, can be formed if and only if for all positive integers m, n, when $nA \leq mB$, then correspondingly $nC \leq mD$. Its intent is to allow for the comparison of ratios when A and B are one kind of thing, and C and D are another kind of thing, while holding to the precept that ratios themselves may only compare things of the same kind. In Greek mathematics, ratios cannot hold between lines and numbers, between finite and infinitesimal magnitudes, or between curved lines and straight lines. The Euclidean tradition treats ratios as relations, different from the things related, and proportions as assertions of similitude (not equality) between ratios.

There is, however, a second, medieval tradition of handling ratios and proportions that originates with Theon, a commentator on Ptolemy's *Almagest*, and is transmitted by Jordanus Nemorarius, Campanus, and Roger Bacon. It associates with each ratio a 'denomination,' that is, a number which gives its size, and in general treats the terms occurring in ratios as well as the ratios themselves uniformly as numbers. Thus ratios are just quotients and the distinction between ratio and term is abolished in so far as they are all numbers.[6] The proportion $A : B :: C : D$ becomes $A/B = C/D$, the equation of two numbers, and so automatically $A \times D = B \times C$. The

---

[5] Ibid., pp. 155–156.

[6] See Edith Sylla, "Compounding Ratios," in E. Mendelsohn, ed., *Transformation and Tradition in the Sciences* (Cambridge: Cambridge University Press, 1984), pp. 11–43.

first tradition governing proportions is invoked here and proves just what Galileo requires; indeed, the second tradition would be unhelpful because in the assertion $A \times D = B \times C$, the product [(time interval) × (displacement interval)] is physically pointless. The really interesting product is [(time interval) × (mean velocity during that time interval)], as will appear below.

In the sequel, Theorem II, Proposition II, and Theorem III, Proposition III, Galileo examines cases involving two particles in uniform motion, and concludes in Theorem IV, Proposition IV: 'If two particles are carried with uniform motion, but each with a different speed, the distances covered by them during unequal intervals of time bear to each other the compound ratio of the speeds and time intervals.'[7] In other words, a precise relationship can be established between any two cases of uniform motion; Galileo formulates it as a proportion: $D_1 : D_2 :: [S_1 : S_2$ compounded with $T_1 : T_2]$. The problem is that "compounding" or finding a product of ratios can only be carried out with continuous ratios, according to the first tradition of handling proportions: to compound the ratios A : B and B : C is to rewrite their combination as $A : C$. However, $S_1 : S_2$ and $T_1 : T_2$ are not continuous, and Galileo is not willing to treat $S_1 : S_2$ and $T_1 : T_2$ as fractions that could simply be multiplied and thus compounded according to the second tradition. Galileo solves the problem by finding the middle term I between D_1 and D_2, which must satisfy the proportions $D_1 : I :: S_1 : S_2$ and $I : D_2 :: T_1 : T_2$: it is the distance the second particle would traverse in the time interval allotted to the first particle. Since we can always find such an I, we can always bring $S_1 : S_2$ and $T_1 : T_2$, and D_1 and D_2 into rational relation. The accompanying diagram, Figure 2, is just a collection of line segments, one each for the speed, time, and distance traversed of body E (resp. A, C and G) and the speed, time, and distance traversed of body F (resp. B, D and L), as well as the seventh line segment, I, which links the two sets of proportions.

The reason why Galileo goes to the trouble of showing that two separate cases of uniform motion can be rationally linked in this manner is because he is going to reduce the uniformly accelerated motion of free fall to a series of cases of uniform motion which then must be brought into rational relation. This is a nice instance of problem reduction, leading a problem about a more complex thing (uniformly accelerated motion) back to a problem about a simpler thing (uniform motion). However, we will see that the

---

[7]Galileo Galilei, op.cit., pp. 157–158.

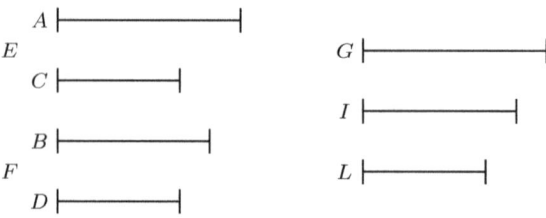

Figure 2.

reduction only works if the intervals in question may be made "as small as one wishes," which of course leads to a highly non-Euclidean employment of the theory of proportions as well as highly non-Euclidean geometric diagrams. Galileo's treatment of the proportions and diagrams later on become carefully ambiguous; and therein lies the innovation.

Theorem I, Proposition I in the section 'Naturally Accelerated Motion' states: "The time in which any space is traversed by a body starting from rest and uniformly accelerated is equal to the time in which that same space would be traversed by the same body moving at a uniform speed whose value is the mean of the highest speed and the speed just before acceleration began."[8] The accompanying figure, Figure 3, has two components, a vertical line CD on the right representing space traversed (again, not just distance but displacement), and a two-dimensional figure $AGIEFB$ on the left, in which AB represents time. The two-dimensional figure reproduces Oresme's diagram that applies the important theorem reached by the logicians at Merton College, Oxford concerning the mean value of a 'uniformly difform form' to uniformly accelerated motion. However, Galileo rotates it by 90° because he is going to apply it even more specifically to the case of free fall, and wants to emphasize its pertinence to the vertical trajectory CD. Koyré points out that the genius of this set of figures is that AB represents not the distance traversed but time, for Galileo (like Oresme) had wrested geometry from the geometer's preoccupation with extension and put it in the service of the temporal processes of mechanics.[9] The left-hand figure represents a

---

[8] Ibid., pp. 173–74.
[9] A. Koyré, "La loi de la chute des corps," in *Etudes galiléennes*, Paris: Hermann,

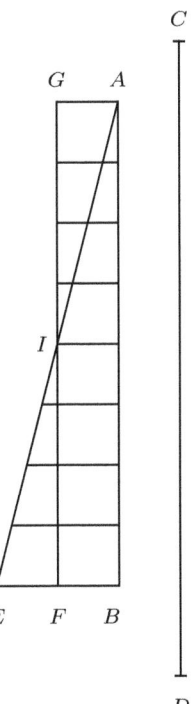

Figure 3.

process like integration with respect to time: the parallels of the triangle AEB perpendicular to AB stand for velocities, and the area of the triangle as a whole, taken to be a summation of instantaneous velocities, therefore represents distance traversed. Distance is then represented in two different ways, as the line segment CD and as the area of the triangle AEB; because the second representation is a two-dimensional figure, it can exhibit the way that uniformly increasing velocity and time are related in the determination of a distance.

1939, pp. 11–46. See also my "Descartes and Galileo: The quantification of time and force," in *Mathématiques et philosophie de l'antiquité à l'age classique: Hommage à Jules Vuillemin*, ed. Roshdi Rashed, Paris: Editions du Centre National de la Recherche Scientifique, 1991, pp. 197–215.

A two-fold representation of distance also occurs in the analysis of free fall given immediately afterwards in Theorem II, Proposition II (Figure 4), but here the right-hand line gains articulation and the left-hand two-dimensional figure loses some: this theorem is about distances, or rather, displacements.

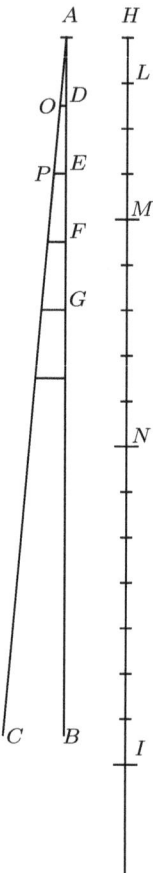

Figure 4.

The theorem states: "The spaces described by a body falling from rest with uniformly accelerated motion are to each other as the squares of the time-intervals employed in traversing these distances."[10] The right-hand figure, the line HI, stands for the spatial trajectory of the falling body, but it is articulated into a sort of ruler, where the intervals representing distances traversed during equal stretches of time, HL, LM, MN, etc., are indicated in terms of unit intervals (by a shorter cross-bar) and in terms of intervals whose lengths form the sequence of odd numbers, 1, 3, 5, 7... (by a longer cross-bar). The unit intervals are intended to be counted as well as measured. In the left-hand figure, AB represents time (divided into equal intervals AD, DE, EF, etc.) with perpendicular instantaneous velocities raised upon it — EP, for example, represents the greatest velocity attained by the falling body in the time interval AE — generating a series of areas which are also a series of similar triangles.

Galileo then considers two cases of uniform motion, and brings them into rational relation, which proves the theorem. He instructs us to draw the line AC at any angle whatsoever to AB and then, given any two equal time intervals AD and DE, to draw parallel lines DO and EP intersecting AC at O and P. He uses the result just proved to show that the distance traversed by a particle falling from rest with uniformly accelerated motion during the time interval AD (resp. AE) is the same as the distance traversed by a particle moving with speed $\frac{1}{2}DO$ (resp. $\frac{1}{2}EP$) during the time interval AD (resp. AE). Thus we know that the ratio $D_1 : D_2$ is the same as the ratio (distance traversed during AD at speed $\frac{1}{2}DO$): (distance traversed during AE at speed $\frac{1}{2}EP$). But what is the latter ratio—how can we bring these two cases of uniform motion into rational relation? The answer is given in Theorem IV, Proposition IV from the section on uniform motion: "the spaces traversed by two particles in uniform motion bear to one another a ratio which is equal to the product of the ratio of the velocities by the ratio of the times."[11] And in this case, because $\triangle ADO$ is similar to $\triangle AEP$, we know that the ratio of $AD : AE$ is equal to the ratio of $\frac{1}{2}DO : \frac{1}{2}EP$, which is just the same as the ratio $DO : EP$; so $[V_1 : V_1$ compounded with $T_1 : T_2]$ is just $[T_1 : T_2$ compounded with $T_1 : T_2]$. Since the ratios only involve the single parameter time, Galileo doesn't mind treating them as numbers and calls the product $[T_1 : T_2]^2$. Thus $D_1 : D_2$ is equal to $[T_1 : T_2]^2$. By the same token, $D_1 : D_2$ is equal to $[V_1 : V_2]^2$, the square of the ratio of the final velocities.

---

[10] Galileo Galilei, op. cit., pp. 174–175.
[11] Ibid., pp. 157–158.

Galileo gains his insight here by combining numerical patterns with geometry in the service of mechanics, as this summary from the immediately following Corollary I indicates: "Hence it is clear that if we take any equal intervals of time whatever, counting from the beginning of the motion, such as AD, DE, EF, FG, in which the spaces HL, LM, MN, NI are traversed, these spaces will bear to one another the same ratio as the series of odd numbers 1, 3, 5, 7; for this is the ratio of the differences of the squares of the lines [which represent time]... While, therefore, during equal intervals of time the velocities increase as the natural numbers, the increments in the distances traversed during these equal time-intervals are to one another as the odd numbers beginning with unity.'[12] Since $1 + 3 = 2^2, 1 + 3 + 5 = 3^2, 1 + 3 + 5 + 7 = 4^2$, and so forth, these sums representing distance will be proportional to the square of the time intervals; and since the time elapsed is proportional to the final velocity, as the similar triangles in the diagram to the left makes clear, the distance fallen will be proportional to the square of the final velocity.

Galileo is now using at least four modes of representation to express his argument: proportions, geometrical figures, numbers, and natural language. He also employs a systematic ambiguity to carry his argument further. By adding Corollary I to Theorem II, Proposition II, he insists on the pertinence of the number theoretical facts just discussed to the analysis of free fall. The reader is thus forced to read the intervals depicted ($AD, DE, EF \ldots$, and then HL, the three intervals of LM, the five intervals of $MN \ldots$) sometimes as units, sometimes as infinitesimals. *There is only one set of diagrams, but the set must be read in two ways.* Reading the intervals as finite allows both for the application of Euclidean results, and for the pertinence of the arithmetical pattern just noted. Reading the intervals as infinitesimal allows for the analysis of accelerated motion. The accompanying text in natural language guides and exploits this double meaning.

Note that in Corollary I, Galileo does not compare the interval-terms directly, but is careful to refer to them in ratios. Even if infinitesimal intervals (instants and points, to use Galileo's vocabulary) are mathematically suspect — as they surely were in the early 17th century — the geometry of the diagrams supports the rationality of holding that ratios between them are "like" the ratios between their finite counterparts. That is, $AD : DE :: AO : OP$ no matter what size the configuration is; or, to

---

[12] Ibid., p. 175–176.

use the other diagram, $HL : LM :: 1 : 3$ no matter what the size of the configuration. Theorem IV, Proposition IV from the section on uniform motion, which is so carefully Euclidean in its reasoning, is here put to highly non-Euclidean use because of its juxtaposition with the systematically ambiguous diagram. When the time intervals AD and AE are read as finite, the application of the theorem that brings disparate cases of uniform motion into relation is direct; when AD and AE are read as infinitesimal (because we may take "any equal intervals of time whatever") the application of the theorem is non-Euclidean because Euclid does not allow infinitesimal terms. But it is *this* application that allows cases of uniform motion to be brought into rational relation with uniformly accelerated motion in a way that Newton can employ when, a generation later, he uses geometry to represent the dynamical processes of the solar system. Read as finite, the triangles are the iconic representations of geometrical figures; read as infinitesimal, the triangles are the symbolic representation of a dynamical process, free fall. What lies before us in Figure 4 is a diagram that must be read in two different ways, as both icon and symbol, and natural language that explains the ambiguous configuration. The same point can be made a propos the left-hand diagram that accompanies Theorem I, Proposition I, (Figure 3) and the diagram borrowed from the Oxford Calculators that adumbrates Corollary I, (Figure 5) in the section 'Naturally Accelerated Motion'.

The centrally important diagram of projectile motion from Theorem I, Proposition I of the *Fourth Day* (Figure 6) also enjoys an ambiguity rich in consequences. The diagram of projectile motion must be compared to the diagram accompanying Theorem I, Proposition I from the section 'Uniform Motion' (Figure 1), the diagram accompanying Theorem II, Proposition II from the section 'Naturally Accelerated Motion' of the *Third Day* (Figure 4), and the diagram of a parabola borrowed from Apollonius just preceding (Figure 7). Significantly, the diagram of projectile motion refers to all of them and conflates and superimposes certain of their elements in instructive ways.[13]

The first thing to note about Figure 6 is that the line *abcde* conflates the two lines in Figure 1, IK representing time and GH representing distance

---

[13]Ibid., pp. 244–257.

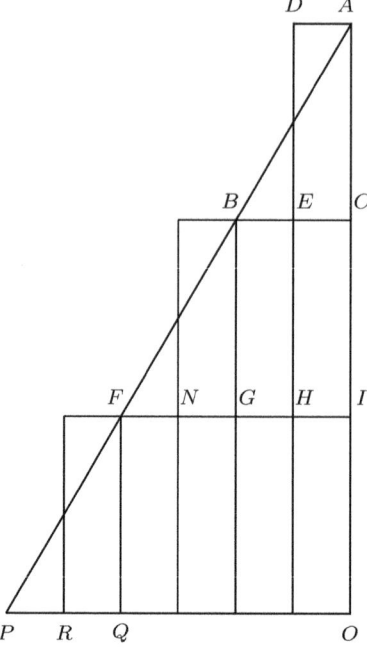

Figure 5.

understood as displacement in uniform motion.[14] The intervening theorems have taught us that it is precisely because line GH represents displacement in uniform motion that it can be merely a line; non-uniform motion requires either a two-dimensional figure or a ruler-with-commentary for its representation. However, in the case of uniform motion, a line suffices and moreover can also serve to represent time, since (as the theorem accompanying Figure 1 states) in such motion the intervals of time elapsed are proportional to the intervals of distance traversed. The line *bogln* is the line HLMNI from Figure 4 divided in just the same proportions. The genius of the diagram is the perpendicular superposition of line *bogln* on *abcde*, which represents the insight that projectile motion is "compounded of two other motions, namely, one uniform and one naturally accelerated."[15] The proof of Theo-

---

[14] Ibid., pp. 248–250.
[15] Ibid., p. 244

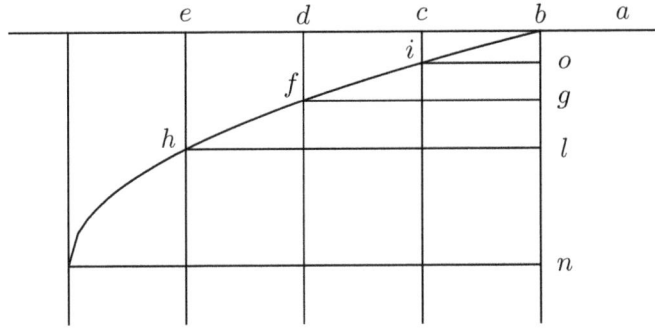

Figure 6.

rem I, Proposition I in the section "The Motion of Projectiles" shows that the rest of the diagram stems from the superposition of Apollonius' construction of the parabola (Figure 7) as the pathway of the moving body: "A projectile which is carried by a uniform horizontal motion compounded with a naturally accelerated vertical motion describes a path which is a semi-parabola."[16]

In order for the reasoning in the proof of Theorem I, Proposition I of the *Fourth Day* to proceed, the line *abcde* must mean both time and distance; it must represent time symbolically in order for the application of the results achieved in the *Third Day* and it must represent distance qua displacement in order for the diagram to make sense as the icon of a trajectory, the movement of a body across a plane in space. The line segments HL, HM, HN from Figure 4, re-labeled *ci, df, eh* in Figure 6, are drawn to represent the vertical displacement at equal intervals of time/displacement at *c, d,* and *e*; *b* is the point taken to represent the beginning of the projectile motion, *cb* is chosen as the "unit" and $cb = dc = de$ and so on. Galileo's exposition of the diagram claims that no matter how *cb* is chosen ("if we take equal time-intervals of any size whatsoever") the curve described is always the same, and it is the semi-parabola, as the results from Apollonius that precede the proof in Theorem I, Proposition I have made clear. Thus the best way to understand projectile motion is "uniform horizontal motion compounded with a naturally accelerated vertical motion" which produces a parabolic

---

[16] Ibid., p. 245.

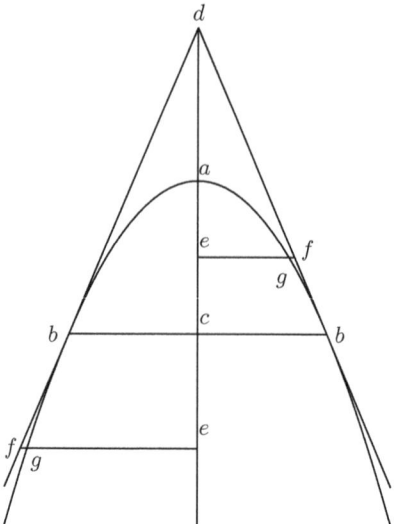

Figure 7.

downward trajectory. Reading cb as a finite interval allows for the application of results of Euclid and Apollonius; reading cb as an infinitesimal allows the diagram to stand as an analysis of accelerated motion. The great diagram that presents projectile motion thus succeeds because of Galileo's inspired handling of controlled ambiguity.

2 Presuppositions About Language and Thought in Mathematics that Stem from Carnap and the Vienna School

The reductionist program set out in Rudolf Carnap's *The Logical Structure [Construction, Aufbau] of the World* can only conclude that Galileo's text should be re-written. Though the programs has long been regarded as inconclusive and ultimately unsuccessful by most Anglophone philosophers of science and mathematics,[17] many of its presuppositions have nonetheless entered into our discourse and thinking. I now return briefly to Carnap's

---

[17]Rudolf Carnap, *The Logical Structure of the World* and *Pseudoproblems in Philosophy*, tr. B. Rolf and A. George, Berkeley: University of California Press, 1967 /69; *Der Logische Aufbau der Welt* was first published in 1928.

classic work, in order to uncover some of these assumptions and question them more closely. Then I will give a brief sketch of what became of Carnap and Hempel's view of scientific and mathematical knowledge in the work of their academic children (Bas van Fraassen, Nancy Cartwright, Margaret Morrison, Ian Hacking) and grandchildren (Robin Hendry, Ursula Klein, myself), a development that will bring us back to the project of this essay in an unexpected way.

Carnap begins his book with the description of his ideal, a "constructional system" that begins with certain fundamental objects/concepts and constructs all other objects/concepts from them. Trying to avoid the antinomy between rationalism and empiricism, he argues that to every concept there belongs one and only one object: "the object and its concept are one and the same."[18] An object/concept is said to be reducible to one or more other objects/concepts if all statements about it can be transformed into statements about the other object/concepts via a constructional definition. This is a "rule of translation which gives a general indication how any propositional function in which a occurs may be transformed into a coextensive propositional function in which a no longer occurs, but only b and c."[19] Then we say that a is logically reducible to b and c. The example that Carnap gives to illustrate his point is pertinent to our discussion:

EXAMPLE 1. The reducibility of fractions to natural numbers is easily understood, and a given statement about certain fractions can easily be transformed into a statement about natural numbers. On the other hand, the construction, for example, of the fraction 2/7, i.e., the indication of a rule through which all statements about 2/7 can be transformed into statements about 2 and 7, is much more complicated. Whitehead and Russell have solved this problem for all mathematical concepts [Princ. Math.]; thus they have produced a "constructional system" of the mathematical concepts.[20]

Earlier in the book, he tells us, "all real numbers, even the irrationals, can be reduced to fractions. Finally, all entities of arithmetic and analysis are reducible to natural numbers."[21] And in the Preface to the Second Edition, he adds that Frege, Russell, and Whitehead had shown that, "through the

---
[18]Ibid, p. 10.
[19]Ibid, p. 61.
[20]Ibid, p. 61.
[21]Ibid, p. 6.

definition of numbers and numerical functions on the basis of purely logical concepts, the entire conceptual structure of mathematics [...] to be part of logic."[22] This highly controversial last statement was, significantly, written in 1961.

The alleged success of Russell and Whitehead at reducing all of mathematics to logic inspires Carnap: "Logistics (symbolic logic) has been advanced by Russell and Whitehead to a point where it provides a theory of relations which allows almost all problems of the pure theory of ordering to be treated without great difficulty." His book is thus designed to "apply the theory of relations to the task of analyzing reality... in order to formulate the logical requirements which must be fulfilled by a constructional system of concepts, to bring into clearer focus the basis of the system, and to demonstrate by actually producing such a system (though part of it is only an outline) that it can be constructed on the indicated basis and within the indicated logical framework."[23] He adds, if this project is successful, it will show "that there is only one domain of objects and therefore only one science."[24]

My aim in this essay is not to evaluate Carnap's overall project, but to note its reductionism (and its optimism about reductionism, even in the face of much evidence to the contrary), and to focus especially on its theory of language. Even philosophers of science and mathematics who reject the constructionist *Aufbau* of the world have accepted many of Carnap's claims about language. For example, he writes that a constructional definition must be "pure," that is, free of unnoticed conceptual elements; and it must be "formally accurate," that is, "it must be neither ambiguous nor empty... it must not designate more than one, but it must designate at least one, object." He notes that in natural language this requirement is difficult to fulfill, but by contrast "this requirement is easily and almost automatically fulfilled when we apply an appropriate symbolism, for example, when we apply the logistic forms for the introduction of classes or relation extensions and for definite descriptions of individuals. It is a fact of logistics that these forms guarantee unequivocalness and logical existence, for they have been created with these desired properties in view."[25]

---

[22] Ibid., p. vii.
[23] Ibid., pp. 7–8.
[24] Ibid., p. 9.
[25] Ibid., p. 154.

In sum, the ideal language for philosophy of science and mathematics, the language in which science and mathematics are to be re-constructed in order to exhibit the real structure of the world, is "the symbolic language of logistics."[26] It does the best job of demonstrating that all objects are reducible to the basic objects: "It is obvious that the value of a constructional system stands or falls with the purity of this reduction, just as the value of an axiomatic exposition of a theory depends upon the purity of the derivation of theorems from axioms. We can best insure the purity of this reduction through the application of an appropriate symbolism."[27] The symbolic language of logistics is allegedly an ideal mode of representation that makes all content explicit; it stands in isomorphic relation to the objects it describes, and that one-one correspondence insures that its definitions are "neither ambiguous nor empty." It is, obviously, symbolic and not iconic; in the ideal limit, it will replace — not merely supplement — natural language. And its successful use in the *Aufbau* of the world will show that there is only one kind of thing: Carnap's choice of basic object is the sense datum, but he also believes that in the end mathematics has no subject matter, having been reduced to the pure formalism of logic.

3 From Syntax to Semantics to Pragmatics in Philosophy of Science and Mathematics

Galileo's account of projectile motion in the *Discorsi* thus appears to fall woefully short of Carnap's ideal. It involves icons and natural language as well as symbols; and many of the modes of representation that it employs, refer ambiguously. Should we commit it to the flames? Carnap's judgment would probably not be so severe; he might suggest we review it as a curiosity, admirable in its time but philosophically inert for us. This is one reason why he was not interested by the history of mathematics. By contrast, I find in my historical case study evidence that is philosophically pertinent in its own right, and that weighs strongly against many of the assumptions made by logical positivists such as Carnap, and in favor of a very different philosophical view of mathematics and indeed of language and logic.

One way to make clear the nature of my quarrel with Carnap and mid-twentieth-century logical positivism is to sketch a philosophical development

---
[26] Ibid., p. 153.
[27] Ibid., p. 154.

that links us, which I understand to be driven by problems articulated but unsolved by his program. It is not unfair to characterize Carnap's project in the *Aufbau* as essentially syntactical, for it reduces content to the sense datum and tries to build everything else into the form. As Robin Hendry sums up: "The logical positivists bequeathed to philosophy of science a characterization of theories as linguistic structures whose content was to be identified in terms of the notion of logical consequence, a notion intimately related to structural features of their formulations in canonical formal languages."[28] If we recall that logical positivists like Carnap and Hempel gave an account of the relations between theory and evidence, between explanation and the *explanandum*, and between the reducing and reduced theory in terms of deductive (only sometimes inductive) relations among sets of sentences in a formal language, the approach appears unrelentingly syntactic.

The philosophical offspring of these mid-twentieth century philosophers concluded that the study of formal languages would not in itself provide a deeper epistemological account of how scientists and mathematicians represent reality. Instead, they turned their attention to "models," and a new school of Anglophone philosophy of science that characterized its approach as semantic began to dominate philosophical debates. Some of these philosophers thought of models in the sense used by logicians, as a structure that satisfies some set of sentences in a meta-language, where what is meant by "structure" is a set of sentences in an object-language; Bas van Fraassen, inter alia, has a rather more conservative view of what constitutes a model. But others urged a broader and richer account. As Robin Hendry observes, "This logical notion is quite different, however, from the sense of 'model' that is at work when we speak of (for instance) the molecular model that we can construct from one of Linus Pauling's kits: here the important relation is not satisfaction but representation."[29] Once the notion of model is investigated along these lines, it becomes clear that different models bring out different aspects of the real systems they model with different degrees of precision and explanatory power. Philosophers who pursue a more broadly semantical philosophy of science, like Nancy Cartwright, Margaret Morrison, Kenneth Schaffner, and Ian Hacking, tend to be interested in the

---

[28] Robin F. Hendry, "Mathematics, Representation, and Molecular Structure," pp. 221-236 in Ursula Klein, ed., *Tools and Modes of Representation in the Laboratory Sciences*, Dordrecht: Kluwer, 2001.

[29] Hendry, op. cit., p. 225.

activity of contemporary scientists, not only as they justify their results in journals but as they discover them in the laboratory and the field.

All the same, as we see in recent works by van Fraassen and Schaffner especially, the adequacy of a scientific theory is characterized in terms of a relation of isomorphism between theory and model. That is, a theory is empirically adequate when a model of the appearances (the quantitative results of experiment) is isomorphic to a (mathematical) sub-model of one of the theory's models. It is often assumed that the best way to think of representation in mathematics is in terms of isomorphism between structures, and this habit can then be simply and appropriately transferred to science. But the philosophical children of the semanticians (and grandchildren of the logical positivists) have concluded otherwise, finding the semantic view of representation problematic not only for philosophy of science but also for philosophy of mathematics. They propose instead a view that encompasses pragmatics as well as syntax and semantics, focusing on the successful posing and solution of problems in a context of use that is, in the end, historical. Thus the program of Carnap – which, like the program of Kant, eschews history – has been transformed into a program that finds it cannot escape history after all.

Ursula Klein concedes that the semantical notion of isomorphism might capture the notion of 'representation of,' or denotation, but "must be supplemented by 'representation as' or meaning." A representation A of an entity B is not merely a denotation of it, but also creatively describes and classifies it as such-and-such. Representation... is not a matter of passive reporting... Rather, representation involves organization, invention, and other kinds of activity."[30] When isomorphism is the central term in the analysis of scientific language, the philosopher assumes that the objects so related are already available and organized in a definitive way. But representation itself may have a role in constituting and organizing the things represented. Klein argues, for example, that the use of Berzelian formulas (the familiar formulas like H_2O used by chemists) was crucial to the swift advance of organic chemistry in the mid-nineteenth century, not because it was "more isomorphic" to experimental patterns recorded in the laboratory than the notations preceding it, but because it enjoyed a useful, meaningful iconicity, ambiguity, and algebraicity.[31]

---

[30] Klein, op. cit., p. viii.
[31] Ursula Klein, *Experiments, Models, Paper Tools: Cultures of Organic Chemistry in*

Hendry reminds us that the notion of isomorphism in fact does not even account very well for denotation in science. One notorious difficulty with first-order predicate logic is that, in most non-trivial cases, a first order theory describes correctly a whole range of models that satisfy it, and cannot by itself pick out the intended model. Hendry argues that the context of natural language in which symbolic language is used makes its reference a determinate, not merely stipulative relation. "The particular historical and material context of a language within which a theoretical discourse is pursued is what endows it with reference, and reference can be passed on to other media (like equations) which become entwined in that discourse."[32] Whereas isomorphism is reflexive and symmetric, representation is irreflexive and asymmetric due to its intentionality.

Moreover, Hendry observes, supposing we find a way to single out the intended model, there are always uninteresting ways of constructing a theory that will stand in the relation of isomorphism to it: the question is then, how to articulate and explain the selection of an *interesting* theory. Klein and Hendry both argue that interesting modes of representation contribute to the advance of scientific knowledge, that is, to success in posing and solving problems. And when we look at the details of their case studies, the representations in those interesting theories turn out to be iconic as well as symbolic, algebraic, and ambiguous; embedded in natural language; and partially constitutive of what they stand for. Thus the semantic reliance on the notion of isomorphism appears misplaced.

My own work in the philosophy of mathematics over the years has led me to the same conclusion about mathematics, an insight missed even by Hendry and Klein who sometimes write as if the semantic approach might work well for mathematics but not for the empirical sciences. Van Fraassen and Schaffner emphatically assume that the relation between theory and thing in mathematics is successfully captured by isomorphism between symbolic meta-language and symbolic object-language, and for the same kind of reason Carnap assumes the syntactic reduction of mathematics to logic is successful. They all suppose that their epistemological account (syntactic or semantic) works well for mathematics and, since mathematics is the language of science, it can be transferred without much difficulty to science itself. My argument in this essay, by contrast, is that an epistemology

---

the Nineteenth Century, Stanford, Ca.: Stanford University Press, 2003, Ch. 1.
[32] Hendry, op. cit., p. 227.

that works properly for mathematics will have to take into account the pragmatic as well as the syntactic and semantical features of representation in mathematics. Focussing on the pragmatic dimension of mathematical language allows us to see the philosophical interest of useful ambiguity in mathematics.

4 Representation Recast in Pragmatic Terms: The Uses of Ambiguity

Carnap wanted to rewrite science (and mathematics) in the language of logic, in order to exhibit the logical structure of reality. Thus for him the role of language is to purify and correct. It should render every inferential step explicit; stand in isomorphic relation with the objects/concepts it refers to; and either reduce content to form or show that there is only one kind of thing. Recall that Carnap's preferred 'thing' is the sense datum, which does not in itself have any content. (The preferred 'thing' of mathematical structuralists is either the point or the empty set, for the same reason, that each lacks content.) Carnap is an enthusiastic champion of the Russellian project of reducing geometry and analysis to arithmetic and arithmetic to logic. He is also interested in the reductionist project of reducing the things of biology and chemistry to the electron (and, he reluctantly adds, the proton). Like that other reductionist champion of conceptual purity, Descartes, Carnap downplays the difficulty of arriving at only one kind of thing. Descartes was left with the *cogito*, the line segment, the particle in inertial motion, and the mechanism; Carnap is left with the sense datum; the proposition; and the electron (and—darn it!—the proton). But the ideal remains.

By contrast, I (along with my pragmatist cohort) regard the role of representation to be the successful discovery and solution of problems about problematic things, heterogeneous things of many different kinds. My exposition of Galileo's analysis of projectile motion in the *Discorsi* is designed to show the fruitful employment of consortia of modes of representation at work in his argument, as well as their inescapable ambiguity, and the same style of exposition can be extended to Newton's use of Galileo's result in the *Principia*. The logicist account of mathematical language makes this useful ambiguity impossible to see, because it tries to eliminate modes of representation that are 'different' — that is, only one mode of representation at a time is countenanced — and it insists that all referring be univocal.

My account of what is going on in I.II.A of *Principia Mathematica* is thus very different from Bertrand Russell's.[33] He claims to show that discourse about numbers can be eliminated by purely logical discourse, replacing "1" and "2" ...et cetera by logical formulas that define "the set of all one-membered sets," "the set of all two-membered sets," and so forth. He thinks that the reductionist task is accomplished by exhibiting a limited isomorphism. What I see on those pages of *Principia Mathematica*, by contrast, is a combination of modes of representation that are very different: Arabic numerals, the notation of predicate logic taken over by Russell from Frege and linearized, and natural language, which provides a setting that forces both numerals and formulas into novel juxtaposition and equivocity. Far from the elimination of arithmetic by logic, what we find there is the emergence of a new problem for logic, that is, how to represent sets of mathematical things in order to exhibit their 'logical complexity.' In the latter role it examines peculiar features of the deductive structure of special theories (in model theory) as well as the fine detail of definitions (in definability theory.) Originally, logic is supposed to be the canon of rules of thought, a formal study of rules of inference that has no subject matter of its own, because it must characterize our thinking about absolutely anything. However, the algebraization of logic in the nineteenth century, in the hands of Boole and De Morgan and then Frege and Russell, gave it a peculiar subject matter, in fact two subject matters: propositions as formal objects, and sets. Russell's reductionist project added a further role for logic, to be a source of canonical representations for other mathematical objects and procedures that registers their 'logical complexity.'

Russell's elaborate formulae for the natural numbers in I.II.A of *Principia Mathematica* are, however, *arithmetically* inert, "lifeless," to use Angus Macintyre's adjective. Russell cannot admit that what he takes to be a symmetric, reflexive isomorphism is an asymmetrically intentional representation designed for the purposes of logic, not number theory. His notation does nothing for the number theorist, since it masks the most important structural feature of numbers, their unique decomposition into prime factors, and lacks the compound periodicity of Arabic numerals that confers an incipient group structure on numbers, suggesting and lending itself to many other such imposed periodicities. Different modes of representation in

---

[33]Bertrand Russell, *Principia Mathematica*, Part II, sec. A, v. 1, Cambridge: Cambridge University Press, 1910, repr. 1963.

mathematics bring out different aspects of the items they aim to explain and precipitate with differing degrees of success and accuracy. Logicians may be interested in Russell's new notation for numbers, but number theorists are not.

Viewed in its role as a canon of representations, first order predicate logic does indeed have an affinity with a special subject matter, depending on how we understand 'aboutness': that subject matter is sets of integers or points on the one hand, and recursively defined well-formed formulae on the other hand. If we collect items into sets (represented by predicates), and then create further sets by projection and complementation (operations represented by the two quantifiers of predicate logic), the increasing 'logical complexity' of the new sets is recorded by the increasing complexity of the well formed formulae. As it turns out, mathematicians did not find it easy to correlate, for example, various hierarchies of point sets that occur in topology with various hierarchies of logical formulae; the reductionist claim became a suite of problems in a research program.[34] In sum, first order predicate logic is not very good at representing the natural numbers themselves; it is better at representing sets of integers or points in one sense and well formed formulae in another, and then exhibiting something useful about them. So understood, it has found some applications in number theory and topology; but that was not what Russell intended.

I hope the foregoing reflection makes it seem an irony that the greatest meta-mathematical results in logic of the last century turned on Gödel's representation of well formed formulae by the natural numbers, a representation whose efficacy stems from their unique prime decomposition. In Gödel's proof, numbers must stand iconically for themselves in order to allow the application of number theoretic results, and symbolically for well formed formulas (which they represent only by convention) to transfer those results to the study of the completeness and incompleteness of logical systems. That study also requires that the logician exploit the constitutive ambiguity of well formed formulas, moving between their iconic interpretation as syntactic items related to other syntactic items by rules of inference, and a symbolic interpretation as representatives of the things of set theory.

---

[34]See my "Two Episodes in the Unification of Logic and Topology," *British Journal for the Philosophy of Science*, 36 (1985), pp. 147–157.

Similarities and Differences Between the Development of Geometry and of Algebra

LADISLAV KVASZ

How far is it possible to reconstruct mathematical reasoning using the tools of logic is one of the central questions discussed in this book. The answer to this question is complicated by the fact that both mathematical reasoning as well as logic underwent great changes in the past, and are still in the process of change. Thus for example before Regiomontanus, Stifel, and Descartes introduced the now standard algebraic notation, formal reasoning based on symbolic manipulations could not exist. Similarly, before Desargues introduced the notion of central projection and enriched the plane with infinitely remote points, any reasoning based on the principle of duality, a standard pattern of reasoning in projective geometry, was unthinkable. Mathematical reasoning is aided by special linguistic tools, and so the historical development of these tools has a fundamental influence on the various patterns of mathematical reasoning. Similarly, on the other side, as long as Aristotelian syllogistic logic was the only logic available, only very few patterns of mathematical reasoning could be recaptured using logical means. After Frege developed the predicate calculus, Turing formalized the notion of computability, and Tarski laid the foundations of logical semantics, much broader realms of mathematical reasoning became accessible to logical reconstruction. Thus the relation of logic to mathematical reasoning is a historical relation.

The aim of this paper is to describe some common patterns of reasoning in geometry and algebra and to try to relate them to some semantic structures. But before turning to the history of mathematics, I would like to introduce some arguments, which stem from the debate on Kant's philosophy of geometry.

1 The debate on Kant's philosophy of geometry

The roots of the contemporary debate on Kant's philosophy of geometry go back to Bertrand Russell, who in his *Introduction to Mathematical Philosophy* formulated a criticism of Kant's position:

> Kant, having observed that the geometers of his day could not prove their theorems by unaided argument, but required an appeal to the figure, invented a theory of mathematical reasoning according to which the inference is never strictly logical, but always requires the support of what is called "intuition". The whole trend of modern mathematics, with its increased pursuit of rigour, has been against this Kantian theory. [Russell, 1919, p. 145]

Thus according to Russell, Kant's mistake was, that he believed in the existence of some forms of reasoning that cannot be captured by logic.

Another important argument against Kantian philosophy was formulated, among others, by Rudolf Carnap who wrote:

> Today it is easy to see the source of Kant's error. It was a failure to realize that there are two essentially different kinds of geometry — one mathematical, the other physical. Mathematical geometry is pure mathematics. In Kantian terms, it is indeed both analytic and a priori... It is simply a deductive system based on certain axioms that do not have to be interpreted by reference to any existing world. ... Mathematical geometry is a theory of logical structure. It is completely independent of scientific investigations; concerned solely with the logical implications of a given set of axioms. Physical geometry, on the other hand, is concerned with the application of pure geometry to the world. Here the terms of Euclidean geometry have their ordinary meaning.... The distinction between the two geometries became especially clear through David Hilbert's famous work on the foundations of geometry. [Carnap, 1966, p181–182]

Nevertheless, at the same time that Carnap published his *Philosophical Foundations of Physics*, Jaakko Hintikka started a series of papers, in which he argued, that Russell's interpretation of the Kantian philosophy of mathematics is misleading. Hintikka identified the point where, according to Kant, mathematical reasoning must make recourse to intuition. This point is the introduction of new objects into the discourse (a proof or a calculation). Thus, Russell is misinterpreting the Kantian position, when he ascribes to Kant the view, that *inference* in mathematics is not strictly logical. Hintikka quotes Kant at several places to show, that in Kant's view not inference, but the introduction of new objects makes the mathematical reasoning synthetic. After the new objects have been introduced, all inferences happen in full accordance with logic.

Some twenty years later Michael Friedman developed Hintikka's arguments further and offered a defence of the Kantian philosophy also against Carnap's criticism. He took from Hintikka the idea, that it is the introduction of new objects into the discourse, which makes mathematical reasoning synthetic. Therefore he compared how new objects are introduced in Euclid's and in Hilbert's treatment of geometry. It turned out, that the main difference between Euclid and Hilbert lies in logic.

> The basis of the modern approach, beginning with Pasch in 1882 and culminating in Hilbert's Foundations of Geometry (1899), is to include an *explicit theory of order*: a theory of the structure and cardinality of the points on a line. ... The presence of some such axioms as 1–6 [axiom of connectedness, of denseness, etc.] is the chief difference between Hilbert's axiomatization and Euclid's. [Friedman, 1985, 464]

The main point is, that for Kant logic was still syllogistic logic, while the theory of order formulated by Pasch or Hilbert required the use of modern quantification theory. Therefore Kant could not recapture notions such as continuity or denseness within his logical framework, and was forced to turn to intuition, which offers us a rather simple image of the continuum. So Kant could not make a distinction between mathematical and physical geometry, simply because the logical tools he had at his disposal, were too weak for a strictly deductive treatment of geometry. Carnap's criticism is therefore

> quite fundamentally unfair to Kant; for, in the first place, Kant's conception of logic is certainly not our modern conception. Our distinction between pure and applied geometry goes hand in hand with our understanding of logic, and this understanding simply did not exist before 1879 when Frege's Begriffsschrift appeared. [Friedman, 1985, 456]

1.1 An argument on the syntactic role of intuitive representations in pre-Fregean mathematics

Friedman's position can be summed up in the form of the following argument: In 1879 Frege introduced a powerful syntactic tool in the form of his quantification theory. This theory makes it possible to formulate the existence of certain objects, like the least upper bound, and to express certain properties, like continuity or denseness. Before Frege, however, the same objects and the same properties had to be represented intuitively, because logic was too weak for this purpose. Thus before Frege intuition was an important component of the construction of mathematical theories and played a fundamental role in mathematical reasoning. Its role was to *compensate the weaknesses of logic*. Only after Frege, when tools for a logical reconstruction of these intuitive representations was available, the recourse to intuition became superfluous. Tools, like diagrams or pictures were degraded to a psychological role and were expelled from the foundations of mathematics.

The logicians, of course, do not like to be reminded that there were times when their beloved logic was a rather weak tool, and they simply project the modern conception of logic on the whole history of mathematics. This prevents them from seeing the important role, which intuitive representations played in mathematical reasoning. This role had nothing to do with heuristics or psychology of invention. Intuitive representations were a fundamental tool, necessary for the construction of the universe of discourse, in times when this universe could not be constructed by the help of logical tools alone.

Besides stressing the importance of intuitive representations in pre-Fregean mathematics, I would also like to stress, that they were not in conflict with logic. Quite on the contrary, intuitive representations were used to fulfil exactly the same roles, which are now fulfilled by logic. Beside the widespread prejudice, that intuitive representations belong to the context of discovery or to the psychology of invention (what is perhaps the case after Frege, but not before) there is another prejudice. Many philosophers believe, that there is some kind of conflict between logic and intuition. But there is no such conflict. As Friedman has shown (the details can be found in his paper), before Frege intuitive representations fulfilled exactly the same roles, which after Frege are fulfilled by logic.

1.2 An argument on the semantic role of intuitive representations in pre-Tarskian mathematics

Even though Friedman's argument against Carnap's criticism of Kant is convincing, it has one fault. It contradicts the historical evidence. Even if *our* modern distinction between pure and applied geometry goes hand in hand with our understanding of logic, it is a matter of fact that a distinction between pure and applied geometry existed already some 50 years before Frege. It appeared for the first time in a letter of Gauss in 1830. A detailed reconstruction of the developments starting with the discovery of non-Euclidean geometry and culminating in Hilbert's *Foundations of Geometry* is presented in a paper of Matthias Schirn on Kant's theory of geometrical knowledge (Schirn 1991). In his paper Schirn offers a rather different account of the transition from the Kantian philosophy of geometry to the modern Hilbertian position. While Friedman stressed the importance of modern quantification theory and the emergence of a theory of order and continuity, Schirn stressed the discovery of the non-Euclidean geometry. Instead of choosing between Friedman's analytic argument and Schirn's historical reconstruction, I will try to reformulate Friedman's argument so, that it would fit the historical evidence, presented by Schirn.

Friedman's argument is based on the analysis of the role of Frege's quantification theory in the creation of modern geometry that means, this argument stresses the syntactic developments in modern logic. Nevertheless, similarly important developments were taking place also on the semantic level. It seems, that the fundamental *logical* innovation that was introduced by the founders on the non-Euclidean geometry, was the introduction of the notion of a *model*. It is of course true, that the first model of non-Euclidean geometry was given by Beltrami in 1868, which is still too late to account for the distinction introduced by Gauss. But the idea of an *interpretation*, which is in many respects very close to the notion of a model, was the basis of the discoveries of the non-Euclidean geometry by Gauss, Bolyai, and Lobachevsky. Let us turn for instance to Lobachevsky, who discovered, that the Euclidean plane can be interpreted on the *horosphere* of a non-Euclidean space (see [Gray, 1979, p. 111–113]). Thus the horosphere is nothing else, but a model of the Euclidean plane in non-Euclidean geometry. Therefore the notion of a model seems to be the fundamental *logical* innovation, which formed the basis of the *historical* developments described by Schirn.

Thus the only thing we need to do with Friedman's argument is to replace syntax by semantics. The introduction of the notion of a model is a deep change of the semantics of a mathematical theory. I believe, that it was precisely this change, that enabled Gauss to see the relation of geometry to physical space in a new way, and so to introduce the fundamental distinction between mathematical and physical geometry. So we can reformulate Friedman's argument in the following form:

> Carnap's criticism is quite fundamentally unfair to Kant; for, in the first place, Kant's conception of semantics is certainly not the same as Gauss' conception. Gauss' distinction between pure and applied geometry goes hand in hand with his understanding of semantics, and this understanding simply did not exist before 1800 when the work of Gauss, Bolyai and Lobachevsky on non-Euclidean geometry started.

In these works, it seems, for the first time the notion of a model was introduced. Of course, we are in pre-Fregean mathematics, thus many aspects of this model are still based on intuitive representations.

2 Towards a reconstruction of the syntactic and semantic development of mathematics

The two arguments, which we extracted from the debate on Kant's philosophy of geometry, shed new light on the role of intuitive representations in the development of mathematical reasoning.

We saw that syntactic constructions, such as the introduction of new objects, or the definition of new predicates and relations, which in contemporary mathematics are done in the framework of Fregean logic, in the pre-Fregean mathematics had to be done using intuitive representations. Therefore we can formulate the *programme of a reconstruction of the syntactic development of mathematics*, which would concentrate on the analysis of the syntactic aspect of these representations. The aim of such a reconstruction would be to see the development of mathematics as a systematic growth of its syntactic tools.

In a similar way the semantic constructions, such as the introduction of a model, which are in contemporary mathematics accomplished using the framework of Cantorian set theory, in the pre-Cantorian mathematics had to be done by the help of intuitive representations. Therefore we can formulate the parallel *programme of a reconstruction of the semantic development of mathematics*, which would concentrate on the analysis of pre-set-theoretical semantic tools. The aim of such a reconstruction would be to see the development of mathematics as a gradual maturation of its semantic structures.

In the development of mathematics we are confronted with a never ceasing process of *linguistic innovations*. To try to create a formal language, in which a particular syntactic or semantic reasoning could be captured in an explicit form, that is what makes mathematics a creative enterprise. As soon as this language is created, logic can come in and turn the particular syntactic or semantic reasoning into an application of some *logical rules*. Even though logic cannot describe the process of the particular linguistic innovations in advance, with hindsight all of the syntactic and semantic arguments can be reduced to logical rules in some sufficiently strong language. I am interested in the understanding of these changes of language, which turn implicit reasoning into logical rules.

2.1 On the programme of a reconstruction of the syntactic development of mathematics

Frege described the syntactic development of mathematics from elementary arithmetic through algebra and mathematical analysis to predicate calculus in his paper *Funktion und Begriff*:

> If we look back from here over the development of arithmetic, we discern an advance from level to level. At first people did calculations with individual numbers, 1, 3, etc.
>
> $$2 + 3 = 5 \qquad 2.3 = 6$$
>
> are theorems of this sort. Then they went on to more general laws that hold good for all numbers. What corresponds to this in symbolism is the transition to the literal notation. A theorem of this sort is
>
> $$(a + b).c = a.c + b.c$$
>
> At this stage they had got to the point of dealing with individual functions; but were not yet using the word, in its mathematical sense, and had not yet formed the conception of what it now stands for. The next higher level was the recognition of general laws about functions, accompanied by the coinage of the

technical term "function". What corresponds to this in symbolism is the introduction of letters like f, F, to indicate functions indefinitely. A theorem of this sort is

$$\frac{dF(x).f(x)}{dx} = F(x).\frac{df(x)}{dx} + f(x).\frac{dF(x)}{dx}$$

Now at this point people had particular second-level functions, but lacked the conception of what we have called second-level functions. By forming that, we make the next step forward. [Frege, 1891]

Nevertheless, it is interesting to notice, that the transitions between the stages described by Frege, were mediated by geometry. Thus for instance between the stage of *calculations with individual numbers* and the stage of *more general laws that hold good for all numbers*, (corresponding to algebraic notation) there was a geometric stage, where line segments of indefinite length were used to represent generality. The length of these segments was not determined (the unit was not chosen) and so the segments could represent any number we wished. They fulfilled in geometric constructions the same function as variables do in algebraic calculations. More details about the reconstruction of the syntactic development of mathematics can be found in my paper [Kvasz, 2000].

2.2 On the programme of a reconstruction of the semantic development of mathematics

In contemporary mathematics the semantic structure of a mathematical theory is described in the framework of set theory. Thus we ascribe to every variable a specific domain, to every function a domain and a range, to every constant an individual. Nevertheless, mathematical theories had a semantic structure also before Cantor developed his set theory. Thus the question arises, what means can we use in order to reconstruct the semantic structure of the mathematical theories of the pre-Cantorian era.

One possibility is to use set theory also in the semantic reconstruction of pre-Cantorian mathematics. But we have seen in the case of the syntactic reconstructions, that the uncritical use of contemporary Fregean logic led to deep misinterpretations of the nature of the mathematical reasoning based on intuitive representations. Both Russell and Carnap misconceived Kant's views precisely because they projected modern logic on the past. Therefore they viewed mathematical reasoning based on intuitive representations as mistaken and misleading, and did not realize, that the recourse to intuition was absolutely necessary precisely because the logical tools of Kant's times were too weak.

I believe, that a similar danger is lurking also behind the attempt to reconstruct the semantics of mathematical theories of the past by simply applying the contemporary set theoretical semantics to these theories. It seems better to look for a tool that would allow us to reconstruct the semantic structure of mathematical theories respecting the important role played by intuitive representations. Wittgenstein's picture theory of meaning from the *Tractatus* can be used as such a tool. This theory, or at least some aspects of it, is appropriate in reconstructing the semantic shifts which occurred in the use of intuitive representations in pre-Fregean and pre-Cantorian mathematics.

While Wittgenstein understood the picture theory of meaning as saying that language is in a sense similar to a picture, I suggest reading it to say, that pictures (intuitive representations contained in mathematical texts) form a kind of language. This makes it possible to use the notions of the picture theory of meaning in the reconstruction of the semantic development of mathematical theories. Of particular interest for our purposes is the notion of *pictorial form* and the distinction between *expressing* and *displaying* ("A picture cannot, however, depict its pictorial form: it displays it." *Tractatus* 5.632). Each figure contained in a mathematical text can be seen as having a specific pictorial form. This form consists of all those aspects of the figure, which do not directly represent anything, but are just displayed. Thus the main objective of our reconstruction is to analyze the pictorial form of the representative means used in mathematics.

Wittgenstein introduced the notion of the pictorial form as one of the central notions of the picture theory of meaning. This notion describes the correspondence between the world and its linguistic representation. The reconstructions of the development of geometry ([Kvasz, 1998]) and of algebra ([Kvasz, 2005]) based on this notion have shown, that the differences between the particular stages in the history of geometry or algebra can be described as differences of the pictorial form of the particular language. Thus the fragment of language, in which a theory, such as projective geometry or polynomial algebra, is formulated, has its own pictorial form. The task of the semantic reconstruction of the development of mathematics can be thus reduced to the task of reconstructing the changes of the pictorial form of their language.

It is known, that Wittgenstein later rejected his picture theory of meaning, because it turned out to be too narrow for the analysis of ordinary language. Nevertheless, I am convinced, that as a tool for a semantic reconstruction of the languages of different mathematical disciplines it is still very useful. The languages of disciplines such as geometry or algebra can be seen as fragments submerged in ordinary language. So even though the picture theory of meaning cannot be used in the analysis of ordinary language, it is very effective in the analysis of the languages of various mathematical disciplines. I will try to present the similarities and differences of the development of algebra and geometry as *semantic* similarities and differences, and I will reconstruct them as the similarities and differences of the particular pictorial forms.

3 On similarities of the semantic development of geometry and algebra

In this section I would like to present some examples from a reconstruction of the semantic development of geometry and of algebra. In this reconstruction I discriminate eight pictorial forms, which in their historical order are:

- *the perspectivist form*
- *the projective form*
- *the coordinative form*
- *the compository form*
- *the interpretative form*
- *the integrative form*
- *the constitutive form*
- *the conceptual form.*

From these eight forms I will present only two, the projective and the integrative form. Each of these two forms will be illustrated by examples both in geometry and in algebra.

3.1 The projective form

The first pictorial form that I would like to discuss is the projective form. The main semantic trick of this form is to construct more representations of the same situation, and then bring these representations into correlation. Thus in projective geometry the same situation is depicted from different points of view, and then these pictures are connected by a central projection. Similarly in algebra the basic idea is to express the same quantity in two or more forms, and then bring these expressions into a relation. So what is here new and what was impossible to express on the previous stages of geometry or algebra are these correspondences or relations. In geometry they have the form of central projections, in algebra the form of substitutions. From the semantic point of view they both have the same function. They make it possible to transform a representation while preserving its reference.

The projective form in geometry
Albrecht Dürer (1471 - 1528) showed us in one of his drawings a method by which it is possible to create a perspectivist painting.

By a similar method the Renaissance painters discovered the principles of perspective. Among other things, they discovered that in order to evoke the illusion of two parallel lines, for instance two opposite sides of a ceiling, they had to draw two convergent lines. They discovered it, but did not know why it was so.

The answer to this, as well as many other questions concerning perspective, was given by projective geometry. Gérard Desargues (1593–1662), the founder of projective geometry, came up with an excellent idea. He *replaced the object with its picture*. So while the painters formulated the problem of perspective as a relation between the picture and reality, Desargues formulated it as a problem of the relation between two pictures. Suppose that we already have a perfect perspective picture of a jug; and let us imagine a painter, who wants to paint the jug using Dürer's procedure. At a moment when he is not paying attention, we can replace the jug by its picture. If the picture is good, the painter should not notice the replacement, and instead of painting a picture of a jug he would start to paint a picture of a picture of the jug.

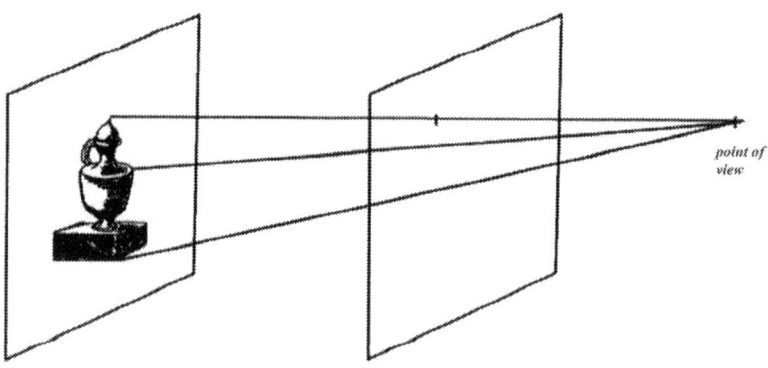

The advantage brought by this idea is that, instead of the relation between a three-dimensional object and its two-dimensional picture we are dealing with a relation between two two-dimensional pictures. After this replacement of the object by its picture, we obtain a central projection of one picture onto the other with its centre in our eye. The *centre of projection* represents the *point of view*.

Before we consider the central projection of any geometrical objects, we have to clarify, what happens with the whole plane, on which these objects are drawn. To make the central projection a one-to-one mapping, Desargues had first of all to supplement both planes with *infinitely remote points*. In this way he created the first technical tool for studying infinity.

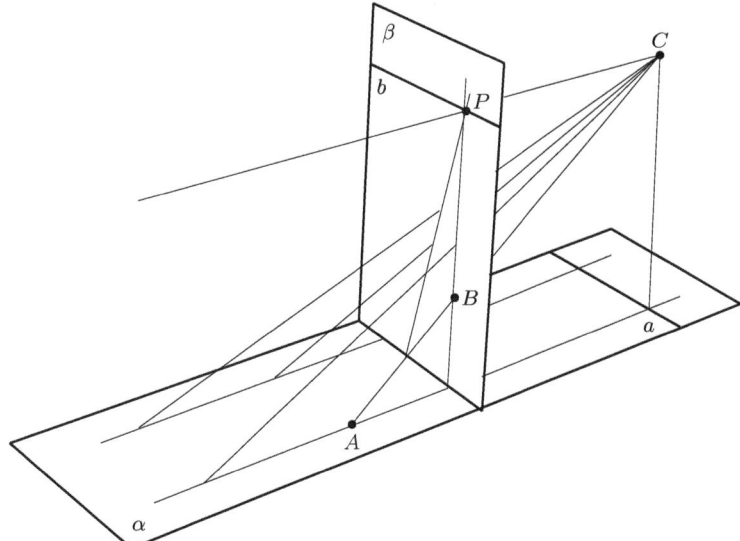

Before leaving our example let us sum up the characteristic aspects of the projective form. The first of them is the creation of the *representation of a representation*. In Dürer's drawing as well as in Desargues' figures not only an object is represented, but it is represented together with its representation. The second aspect is the explicit representation of the *point of view*, be it the eye of the painter in Dürer's drawing or the centre of projection in the figures of projective geometry. The third aspect is a representation of the infinitely remote points, which we call *ideal objects*.

The projective form in algebra

The solution of a cubic equation was first published by Girolamo Cardano (1501 - 1576) in his famous *Artis Magnae sive de Regulis Algebraicis* in 1545. In books on history of mathematics the central idea of Cardano's solution of the equation of the type

$$x^3 + bx = c$$

is interpreted as the substitution

(1) $\quad x = \sqrt[3]{u} - \sqrt[3]{v}$.

Before the Italian school of algebraists of the 16th century, to which Cardano belonged, the mathematicians used only one expression for the unknown. It was not x (this convention comes from Descartes), but r, the first letter of the Latin word *res*. For the convenience of the reader we shall indicate the unknown by x. The substitution (1) is a great innovation, because it introduces a new representation for the unknown, and so the formula (1) itself can be seen as analogous to Dürer's drawing, as a *representation of a representation*. It represents the same thing, namely the unknown, twice. First it represents the unknown using the letter x (which can be seen as an analogy to the jug in Dürer's drawing) and then as $\sqrt[3]{u} - \sqrt[3]{v}$ (which is an analogy to the picture of the jug in the drawing). Further, there is the sign =, which represents the relation between these two expressions. In Dürer's drawing the eye of the painter (becoming the centre of projection in Desargues) was the point, which founded the sameness of the jug and its image in the representation. Therefore the sign = is an analogy of the *point of view* in algebra. As Frege has already shown, the sign = does not express any relation between things, therefore it does not belong to the expressions of the language, representing something from the domain of the theory. It can be rather seen as an aspect of the pictorial form, an aspect analogous to the centre of projection in the figures of projective geometry. The third aspect, which underlines the analogy between Cardano and Desargues, is the discovery of the *casus irreducibilis*, which finally led to the introduction of the complex numbers. Complex numbers are, in our view, *ideal objects*, just as were the infinitely remote points in projective geometry. Their introduction or, in other words, an extension of the domain of the theory, is another typical aspect of this pictorial form.

We see here an analogy with geometry. The new pictorial form in algebra had all the features, that we found in geometry: *a representation of a representation, a point of view*, and *the introduction of ideal objects*.

3.2 Integrative form

The second pictorial form, which I would like to discuss, is the integrative form. The detail can be found in my paper Kvasz 1998, so I will concentrate on the main idea. In geometry the integrative form corresponds to Klein's

Erlanger program, in algebra it corresponds to *Galois' theory*. These two theories have a remarkably similar semantic structure. The main semantic trick in both of them consists in the separation of the structure (the metric structure of the plane in geometry and the structure of solvability of an equation in a field in algebra) from the ontological basis of this structure (the plane or the particular field). The ontological basis of the theory becomes in this way a neutral medium, and the structure, after being separated from this medium, gets the form of a symmetry group (the group of transformations of the particular geometry, or the group of automorphisms of the particular field). The gain obtained by this separation of the geometric or algebraic structures from their ontological basis is the possibility of comparing different such structures with each other. At the previous stages of development it was impossible to compare different structures, because their ontological fundaments were incompatible. Only due to separating the structures from their ontological foundations Klein became able to compare the structures of symmetries of the Euclidean and non-Euclidean geometries and Galois was able to compare the structures of symmetries of solvable and insolvable equations.

The integrative form in geometry

In geometry the integrative form is connected with the works of Arthur Cayley (1821 - 1895) and Felix Klein (1849 - 1925), and especially with the *Erlanger program* (1872). Before Cayley and Klein geometry was generally understood as something based on visual intuition. Therefore, although the non-Euclidean geometries were discovered one generation earlier, the Euclidean and non-Euclidean geometries were thought to be exclusive. Either we see the world in the Euclidean way, and then the non-Euclidean geometry is excluded, or we see the world in the non-Euclidean way, and then the Euclidean geometry is false. We cannot see the world simultaneously in the Euclidean and the non-Euclidean way. This mutual exclusivity of the two geometries was preserved also in Beltrami's model of the non-Euclidean plane, because from the point of view of visual representation Beltrami used an ordinary Euclidean plane. Therefore his model is based on the Euclidean spatial intuition. Beltrami only speaks about objects, which he intuits in a Euclidean way using a non-Euclidean vocabulary.

Klein (taking up the ideas of Cayley) changed this situation in a radical way. For Klein geometry is based not on visual intuition as such, but on transformations. Geometry is the study of the invariants of transformation groups. When we take geometry as a form of visual intuition, Euclidean and non-Euclidean geometries exclude each other. But when we understand geometry in the Kleinian way, that is as systems of invariants of transformation groups, the two geometries become compatible. They even exist together as subgroups of the projective group. Therefore we understand Klein's move as a change of the pictorial form of the geometrical language. While using the languages based on the previous pictorial form it was impossible to study the relations between the various geometries, because they excluded each other. In the *integrative form*, the pictorial form introduced into geometry by Klein, all the different geometries are integrated into one general representation, and so it becomes possible to study their mutual relations.

The integrative form in algebra

I am convinced, that the integrative form is also the foundation of the Galois theory. As long as we understand the process of solving of an algebraic equation as a procedure of manipulation with algebraic formulas, all such procedures are mutually exclusive. If we make as a step of this process a specific algebraic transformation (an addition, a multiplication, a root extraction), all other alternatives are excluded. At one moment we can make only one algebraic operation, and if we decide to make one, all the others are excluded. To prove the insolubility of the quintic equations first of all this exclusiveness of the different algebraic manipulations has to be overcome. This was done by Evariste Galois (1811 - 1832), who introduced the notion of a splitting field and interpreted the process of solving an equation as a factorization of the so-called Galois group of this field. The splitting field of an algebraic equation is analogous to the projective plane in geometry. They both form a neutral background, which makes the introduction of a specific structure possible. And in both cases this structure has the form of a group. In algebra it is the group of automorphisms of the splitting field, in geometry it is the transformation group of the projective plane. And in both cases the group makes it possible to incorporate specific structures, which originally excluded each other, into one representation. In algebra different normal subgroups of the Galois group represent different ways of solving the equation, while in geometry the different subgroups of the projective group represent different geometries. The fundamental breakthrough in algebra, consisting in proving that the *alternating group* $\mathbf{A}_5$ has no nontrivial normal subgroups, and so the general equation of the fifth degree is insoluble, is thus based on the introduction of a language with a new pictorial form, the *integrative form of the algebraic language*.

4 The various aspects of the pictorial form

I illustrated my approach by examples of the projective and the integrative forms. I hope that these examples were able at least to convey an impression of what I mean by the pictorial form. Now we can pass on to a closer examination of this notion. We can discriminate six aspects of a pictorial form:

> *the epistemic subject of the language,*
> *the horizon of the language,*
> *the individua of the language,*
> *the fundamental categories of the language,*
> *the ideal objects of the language,*
> *the background of the language.*

I believe, that these aspects are formal, i.e. they have no factual meaning. Let me explain what I mean by this on the example of the horizon. If we take a perspectivist painting of a landscape, we can clearly recognize a line, which is called the horizon. Nevertheless, if we went out in the countryside, represented by the painting, to the place of the alleged horizon, we would find nothing particular there. And the painter, when painting his landscape, did not paint the horizon by a stroke of his brush. He painted only houses, trees, hills, and at the end the horizon was there. This is the meaning of Wittgenstein's words: "A picture cannot, depict its pictorial form: it displays it". The painting does not depict the horizon; it displays it. The horizon is an aspect of the pictorial form and that means that it cannot be empirically determined. Despite the fact, that in the picture the horizon can be clearly determined, in the world represented by the picture there is no object corresponding to it.

It is interesting, that the languages of mathematical theories are full of such non-denotative expressions. Take for instance the zero or the unit in different algebraic structure, the negative or the complex numbers. The purpose of such non-denotative expressions is to connect the subject (the user of the language) with this world (the universe of the language). This connection happens on three levels.

4.1 The incorporation of the subject into the world

The first function of the pictorial form is to incorporate the epistemic subject into the world. This is achieved with the help of the point of view and of the horizon. The point of view (in projective geometry it was the centre of projection, in algebra the number zero) indicates the position of the subject, from the viewpoint of which the theory is formulated. The point of view thus incorporates the speaker into the world; it constitutes the *identity of the subject* in the universe of the language.

The horizon (in projective geometry the vanishing line, in algebra the unit) coordinates the world and the epistemic subject. When it fixes the basic directions (the vanishing line determines the horizontal plane and in this way it determines the directions upwards and downwards; the unit determines the positive direction of the number line, thus discriminating the increasing from the decreasing), it constitutes the *situatedness of the subject* in the universe of the language.

4.2 The structuring of the world from the point of view of the subject

The second function of the pictorial form is to structure the world from the point of view of the epistemic subject. This is achieved with the help of the introduction of the individua and of the fundamental categories of the language. To determine the individua means to identify objects in the world that are in a sense analogous to the subject; objects that the epistemic subject can refer to in a fixed way. Not accidentally the term "body", by which individua are designated in classical mechanics, has its roots in the Old English term for corpse. A body is something analogous to our human body, something we can refer to in a corporeal way. *Individuality* is a fundamental attribute of the subject. The subject encounters himself as an individuum, and projecting his individuality to certain objects, he constitutes them as individua of the language. This is why the determination of the individua cannot be an empirical question, which could be decided independently from the particular pictorial form. Thus for instance in projective geometry, incorporating the infinitely remote points into the language, the two parts of a hyperbola are considered as forming a single curve, i.e. an individuum. Earlier it was natural to consider them as two different objects, i.e. as two individua. On a more abstract level even the three conic sections, the ellipse, the parabola, and the hyperbola, can be seen as three different positions of the same object, i.e. as a unique individuum.

The next step to make, after the division of the homogenous continuum of being into discrete individua is to introduce different *similarity relations* into the world. Congruence, similarity or affinity are introduced into the set of all geometrical figures; similarly different congruence relations are introduced into algebra. In this way language introduces certain structures into the world, and so it makes the world more familiar for the subject.

4.3 The homogenization of the world

The third function of the pictorial form — besides the incorporation of the subject and the structuring of the world — is to introduce a neutral, homogenous background into the world. This helps the subject to find his orientation in the world as a whole. Typical elements of this kind are the different kinds of space in geometry (projective space, affine space, metric space, topological space), the different number systems in algebra (natural numbers, real numbers, fields, integral domains, rings). These notions do not refer to anything real, so we could possibly avoid using them. But it is obvious, that this would only make our language more cumbersome. Thus it is useful to add these formal objects to our language, which can help us to find our *orientation* in the world more easily.

On this homogenous background we build different schemes. Nevertheless, it might happen, that the world is not in accordance with the schemes, we use. Then it is useful to add some ideal elements to the world, just as the infinitely remote points are added to space in geometry, or the complex numbers are added to the number system in algebra. These again are expressions of the language, having no real denotation and therefore we include them into the pictorial form. But if we add them to the language, the world becomes more *transparent*. Schemes, which originally had only a restricted validity, become universal. After having added the infinitely remote points to the plane any two straight lines will have an intersection. Thus in the proofs it is not necessary to distinguish the different cases depending on the intersection of lines. Similarly in algebra after having added the imaginary numbers to the number realm every number will have a square root, and so it is not necessary to distinguish the different cases, depending on whether an expression is positive or not. All the schemes start to work much more smoothly, the world becomes more transparent.

We have seen, that the purpose of the respective aspects of the pictorial form is to situate the epistemic subject in the universe of the language, to structure this universe in a way comprehensible for the subject, and to provide means for a better orientation in the so structured universe. The aspects of the pictorial form, which constitute the identity, situatedness, individuality, similarity, orientation, and transparency, are not factual. The subject does not belong to the world (*Tractatus* 5.632: *The subject does not belong to the world: rather, it is a limit of the world*), the horizon does not denote anything factual, there are no fix individua, the space does not really exist, not to speak about the ideal elements. On the other side, we cannot deny, that all these aspects of language are useful, they help us to speak about the world in a more comprehensible way.

5 On differences between the semantic development of geometry and algebra

It is rather surprising that the languages of such different disciplines as geometry and algebra have so much in common. The evolution of these disciplines can be characterized as the development of the pictorial form of their languages. This is quite natural, because linguistic expressions with direct or indirect denotations are bound by this relation of denotation, and so cannot change their meanings. On the other side the various aspects of the pictorial form, because they do not denote anything real in the world, are to a great extent free. They are bound only by their mutual interdependence. If then the development of knowledge requires a change of language, the various aspects of the pictorial form offer enough space for innovation.

5.1 The omission of some pictorial forms

Perhaps the most important difference between the reconstruction of the development of geometry and of algebra concerns the *coordinative* and the *compositive forms*. While these forms are clearly present in the development of algebra, in the reconstruction of the development of geometry I could not find them.

Thus one kind of differences between the development of geometry and algebra may consist in the fact, that some of the eight above mentioned pictorial forms can play no role in the development of one of these disciplines, while it has an important role in the development of the other. This sheds some light on our method of reconstruction of the semantic development of mathematical theories. The list of eight pictorial forms, that I presented earlier, represents a succession of different possibilities, of how the correspondence between the intended domain of the theory and its expressions can be arranged. These forms are ordered successively according their growing complexity. If in the use of the theory some difficulties appear, one of the possibilities of how to solve them is to turn to a language with a more complex pictorial form, a form that is able to represent more complex situations. Therefore the most natural thing to do is to make one step in the succession of forms presented in our list. Nevertheless, it may happen, that the next pictorial form, that is the language with the smallest possible complication of the semantic structure, is not appropriate for the solution of the particular problems, which the theory encountered. So it may be necessary to make a more radical change, what will be manifested in omitting one or two stages in our list and turning to some more elaborate pictorial form. It seems that this happened in the development of non-Euclidean geometry, when the coordinative form was of no use for the problems the founders of these new geometries encountered, and so they turned directly to the interpretative form. Therefore our list offers a maximal system of stages, a complete (at least I hope so) system of possible arrangements of the semantic correspondence of the expressions of the language with its intended universe, and from this list the actual development can select the appropriate ones.

5.2 The differences of intuitive representations used in geometry and in algebra

An interesting aspect of the reconstruction of the development of geometry was a regular alternation of *implicit* and *explicit* forms. This was possible, because in geometry each pictorial form exists in these two clearly distinct versions. The situation in algebra is much more complicated. If we look at the development of the algebraic symbolism, we see that it is a rather slow and gradual process, stretching from Regiomontanus to Descartes, over more than two centuries. The dynamics here, instead of a change from an implicit version of the pictorial form to the explicit version (as was the case in geometry) consists in a slow process of reification of the expressions of the language. The world of geometry is opened as a whole to our sight, and therefore any change of it must happen at once, as a *Gestalt switch*.

In algebra, by contrast, the world that is given to us is only fragmentary, we know only some of its "places" where we have "fumbled around" (in our calculations), we know only a few "tricks" which we have found (as the substitution $x = \sqrt[3]{u} - \sqrt[3]{v}$). Thus in algebra the emergence of a new pictorial form happens slowly and gradually, and does not resemble a Gestalt switch.

The contrast between the world opened to our sight and the world constituted in the process of reification of linguistic expressions enables us to explain another peculiarity of algebraic texts. Let us take Cardano's *Ars Magna*. When we take up this book, we discover that it contains fragments belonging to different pictorial forms, fragments having very different semantic structure. It contains the famous rules for the solution of cubic equations. These rules are formulated in ordinary language, which is characteristic of the *perspectivist form*. As we have shown, these rules were derived with the help of a substitution, which is a typical feature of the *projective form*. Further, the book contains the famous *casus irreducibilis*, what is the germ of a new, *coordinative form*, the form in which the complex numbers will be incorporated into algebra. Thus it seems that an algebraic text may contain different fragments, belonging to three different pictorial forms. In geometry such a coexistence of fragments belonging to different pictorial forms is inconceivable. It is impossible to have a picture, which is partly Euclidean, and partly non-Euclidean. In geometry the world is disclosed as a whole, and thus all its parts must fit together.

Acknowledgements

I would like to thank to Donald Gillies and Eberhard Knobloch for their criticism of the previous versions of this paper. I would like to express my gratitude to the *Alexander von Humboldt Foundation* for a scholarship, which made it possible to write this paper. The financial support of the grant VEGA 1/0223/03 is acknowledged.

BIBLIOGRAPHY

[Carnap, 1966] Carnap, R. *Philosophical Foundations of Physics*, Basic Books, New York, 1966.

[Frege, 1891] Frege, G. *Funktion und Begriff* , English translation in *Translations from the Philosophical Writings of Gottlob Frege*, edited by P. Geach and M. Black, Basil Blackwell, Oxford 1952, pp. 21–42.

[Friedman, 1985] Friedman, M. Kant's theory of geometry, in *The Philosophical Review*, Vol. **94**, pp. 456–506.

[Friedman, 1992] Friedman, M. *Kant and the Exact Sciences*, Harvard University Press, Cambridge.
[Gillies, 1992a] Gillies, D. (ed.) *Revolutions in Mathematics*, Clarendon Press, Oxford, 1992.
[Gillies, 1992b] Gillies, D. *The Fregean revolution in Logic* , in [Gillies, 1992a].
[Gray, 1979] Gray, J. *Ideas of Space Euclidean, Non-Euclidean, and Relativistic*, Clarendon Press, Oxford, 1979.
[Hintikka, 1965] Hintikka, J. Kant's new method of thought and his theory of Mathematics. In *Ajatus*, Vol. **27**, pp. 37–43, reprinted in *Knowledge and the Known, Modern Essays*, Reidel, pp. 126–134, 1965.
[Hintikka, 1966a] Hintikka, J. Kant vindicated, P. Weingartner (ed.), *Deskription, Analytizität, und Existenz*, Salzburg, Pustet, pp. 234–253, 1966.
[Hintikka, 1966b] Hintikka, J. Kant and the tradition of analysis , P. Weingartner (ed.), *Deskription, Analytizität, und Existenz*, Salzburg, Pustet, pp. 254–272, 1966.
[Kvasz, 1998] Kvasz, L. History of geometry and the development of the form of its language, *Synthese*, Vol. **116**, pp. 141–186, 1998.
[Kvasz, 2000] Kvasz, L. Changes of language in the development of mathematics, *Philosophia Mathematica*, Vol. **8**, pp. 47–83, 2000.
[Kvasz, 2005] Kvasz, L. The history of algebra and the development of the form of its language, sent to *Philosophia Mathematica*, to appear 2005.
[Russell, 1919] Russell, B. *Introduction to Mathematical Philosophy*, Routlege, London, 1993.
[Schirn, 1991] Schirn, M. *Kants Theorie der geometrischen Erkenntnis und die nichteuklidische Geometrie*, in *Kant Studien*, Vol. **82**, pp. 1–28, 1991.
[Scholz, 1990] Scholz, E. (ed.) *Geschichte der Algebra*, Wissenschaftsverlag, Mannheim, 1990.
[Wittgenstein, 1921] Wittgenstein, L. *Tractatus logico-philosophicus* , Frankfurt am Main: Suhrkamp, 1964.

Reflections on the Proliferous Growth of Mathematical Concepts and Tools: Some Case Histories from Mathematicians' Workshops

YEHUDA RAV

1 Introduction

The interplay between theory and experimentation/observation is the key to the growth of scientific knowledge. For the empirical sciences, the development of appropriate technologies is essential in furnishing the material tools for experiments, thus forming an integral part of the scientific enterprise. Theories suggest experiments that in turn substantiate or invalidate theoretical claims; experimental/observational results call for theoretical explanations. Experiments require a supportive technology, tools have to be invented and developed, and there is a constant flux between these three supportive poles in the dynamics of scientific investigations as schematically represented in the diagram below.

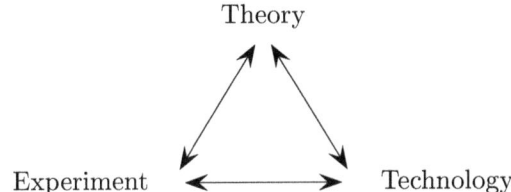

How does this scheme representing the dynamics of the growth of knowledge in empirical sciences compare with the development of mathematical knowledge?

In a nutshell, in the case of mathematics, the three apices in the above diagram *coalesce* to a single point. In the creative work of the mathematician, the parts of the theoretician, toolmaker, and experimentalist[1] are organically and inseparably united. Mathematicians are problem solvers.[2] The problems may originate from external, scientific needs, yet frequently are just internally generated, but not necessarily independent in the long run from indirect external inputs.[3]

The solution of problems calls for the invention of *conceptual tools* that in turn solve or reduce the problem at hand, only to suggest further problems. Tools, which have been developed for attacking a particular problem, are then added to the mathematician's tool shelf for possible other uses and further developments. Tools are made with tools, "pliers are made with pliers" (says the Hebrew proverb).[4] In this dialectic process, from problem to tools and then to new problems, mathematical knowledge *proliferates*, with new buds being continually formed. Mathematics grows like a tree, through (a) the dynamic interaction of internal ontogenetic factors—read, internal theoretical developments; (b) nutriments supplied by the environment— read, demands and catalyzing stimulation coming from the various sciences;

---

[1] Traditionally, experiments in mathematics are *thought experiments* for searching connections and possible deductive paths. A novelty of late is the use of computers in exploring and formulating conjectures or as an aid in searching for proofs. Testing conjectures by computers has something of an experimental procedure, perhaps best being referred to as a quasi-empirical procedure. This being said, none of these make mathematics a quasi-empirical science *in its totality*.

[2] It has cogently been argued by [Cellucci, 2000] that "the primary activity of mathematics is not theory construction but problem solution, and that a satisfactory philosophical account of the growth of mathematical knowledge, one that can hope to account for the rationality of its discovery process, must take this priority into account". (Cited from the Introduction by the Editors on p. xxii in the volume in which the article has appeared).

[3] The relative importance of "external inputs" is frequently debated in the philosophy of mathematics, giving rise in variously shaded formulations to an empiricist philosophy of mathematics. For a recent defense of this position, with due attention to the history of empiricisms in the philosophy of mathematics, see [Gillies, 2000].

[4] From the same vantage point, Marquis (1997) writes: "Mathematicians (almost) *literally* talk about instruments. They use this expression in its primary sense: the instruments are certain constructions applied to certain 'objects' in a given context in order to obtain useful information. The analogy with instruments in the empirical or factual sciences is surprisingly close". (Emphasis in original). See also Marquis [1999].

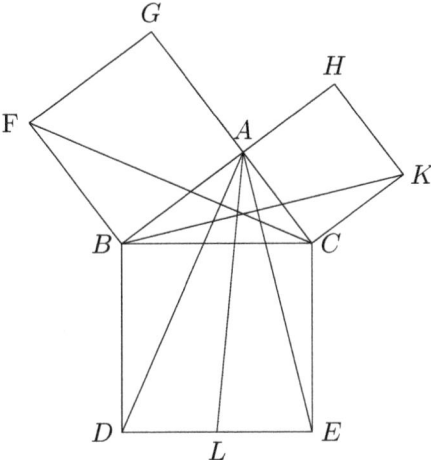

Figure 1.

(c) being planted in the right soil—read, culture. The contribution to the growth of mathematical knowledge of each of these factors varies from epoch to epoch and from branch to branch. I hope to bring out, at least implicitly, the interplay of these various growth factors through the case histories to be presented in the sequel, with the unifying thread leading from the Pythagorean relation, via the avatars of intermediate developments, to the use of the Riemannian metric, *qua* tool, in Einstein's theory of general relativity.

2 The Pythagorean theorem: Euclid's proof versus "cut-and-paste" demonstrations

We all recall, for good or for bad, our first encounter with the Pythagorean theorem in high school, as stated and proven in Euclid's Elements. Prop. 47, Book I says: "In any right-angled triangle, the square which is described on the side subtending the right angle is equal to the squares described on the sides which contain the right angle". Euclid's proof proceeds as follows (see Fig.1).

Dropping a perpendicular from A to the base of the square BE, meeting it on the line DE at the point, call it L, Euclid argues that $\triangle ABD$ is congruent to $\triangle FBC$, that rectangle $BL = 2\triangle ABD$, and rectangle $GB = 2\triangle FBC$,

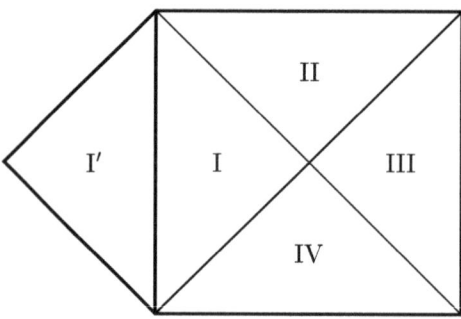

Figure 2. (after Michaels [1978, p. 121])

using previous propositions. Thus, rectangle BL = square GB. Likewise, rectangle CL = square AK.

Euclid's proof is a *tour de force* to fit a preset methodology of a purely geometric, deductive argument. Notice that Euclid speaks of areas, not even of the length of the sides of right triangles. As it stands, Euclid's proof hides the heuristic path to the discovery and a more intuitive proof of the Pythagorean theorem. However, there were also other traditions in antiquity in which axiomatics played no role and where the demonstrations of geometric relations consisted simply of *démonstráre*, to point out, to show. Indeed, to fix our attention just on the Pythagorean theorem, the relation between the lengths of the sides in a right-angle triangle has been discovered repeatedly in different societies and *shown* to be valid in a very concrete way. Below are three such examples, based on reconstructions of transmitted documents. One set of examples comes from India, the other from China, which typify the "cut-and-paste" methodology of geometric demonstrations.

Figures 2 and 3 come from two different Sulvasûtras, ["Manuals of the cord"], which embody various texts of Vedic sacral geometry.[5] Both Sûtras treat the special case of isosceles triangles, by apparently having constructed actual physical models and argued upon appropriate folding. Thus, one "sees" that in Fig. 2, the area of the small square $I'+I$ equals $1/2$ the area of large square, hence in the right-triangle I', 2 (side in I')^2 = (hypotenuse)^2.

---

[5]The dating of these texts is uncertain; some scholars place them to have originated in an oral tradition way back to the 16th century BCE. The written documents seem now to be placed between 500 BCE and 200 BCE.

Reflections on the Proliferous Growth of Mathematical Concepts 53

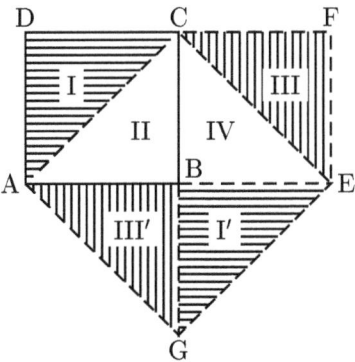

Figure 3. (after Müller [1930, p. 176])

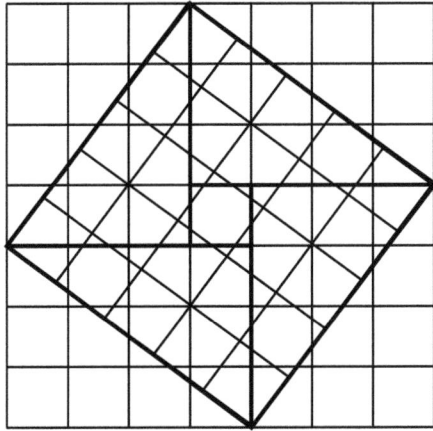

Figure 4. (after Needham [1959, vol.3, p.22])

As to Fig. 3, again by a folding argument, it follows that $2\,(AB)^2 = (AB)^2 + (BE)^2 = (I+II) + (III+IV) = (I'+II) + (III'+IV) = (AC)^2$.

Fig. 4 comes from a printed image on cloth in a text of the *Chou Pei Suan Ching*.[6] Notice that Fig. 4 treats the case of a 3;4;5 triangle. These examples illustrate the heuristic stage of early geometric ways of discovering and actual showing by *folding* that which later, in the more abstract deductive setting of Euclid, resulted in propositions about *congruences*.[7]

A related problem in connection with the Pythagorean theorem concerns the *arithmetic* problem of finding *Pythagorean triples*, i.e., whole numbers x, y, and z for which $x^2 + y^2 = z^2$. According to Proclus (c. 410–485), both Plato and the Pythagoreans discovered a method for obtaining infinitely many such triplets. Needless to say, no mention of this problem in Euclid's Elements, as it does not fit into Euclid's rigid geometric methodology.

3 The analytic avatars of the Pythagorean theorem

Let us move to the 17th and 18th centuries to examine how the Pythagorean theorem was transformed with the new tools of analytic geometry and the calculus, opening the road to some far reaching novel developments.

Starting with Galileo (1564–1642), Kepler (1571–1630), and through the monumental contributions by Newton (1642–1727), a new spirit emerged: to study nature mathematically. Under this *external impetus,* coupled with important *internal developments,* such as the perfection of symbolic notations and advances in algebra, the stage was set for an accelerated growth of mathematical knowledge.

As is well known, in the spirit of developing *methods* for philosophical inquiries, Descartes (1596–1650) was led to invent a new method for geometric investigations, namely, analytic geometry. Typically, his "De la Géometrie" was added just as an appendix to his major treatise "Discours de la Méthode", published in 1637. "Actually", writes Kline [1972, p. 322], "it was the use of algebra in geometry that [Descartes] undertook to exploit. He saw fully the power of algebra and its superiority over the Greek geometric methods in providing a broad methodology. He also stressed the

---

[6] Again, the dating is uncertain, but could be as late as the third century CE.

[7] The issue of proofs by pictures has been subjected to a recent debate, with arguments for and against. See the discussion in Dove [2002] and the literature cited therein. In connection with the ancient picture discovery/proof of the Pythagorean theorem, see in particular p. 337 (op. cit.).

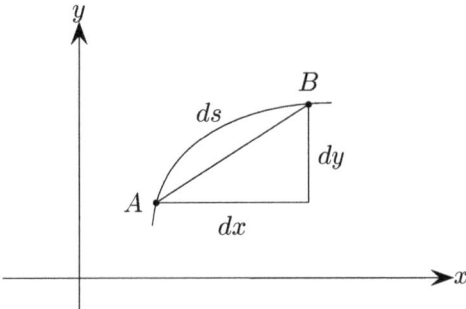

Figure 5.

generality of algebra and its value in mechanizing the reasoning process and minimizing the work in solving problems. He saw its potential as a universal science of method." And further: "Indeed, the enormous power mathematics developed from the 17th century on must be attributed, to a very large extent, to coordinate geometry".

In terms of the "machinery" of analytic geometry, using rectangular coordinates, the distance d between two points $A(x_1, y_1)$ and $B(x_2, y_2)$ can now be *computed* in terms of the coordinates of the respective points by a straightforward use of the Pythagorean theorem:

$$d^2 = (x_1 - x_2)^2 + (y_1 - y_2)^2$$

Similarly, with a third term added in three dimensions. The next step, however, introduces the "infinitesimal" Pythagorean relation in terms of the calculus in the manner of the 18th century.

If A and B are two points on a curve as in Fig. 5, and if A is *infinitely close* to B, so it was argued heuristically, then the *arc ds* joining A to B on the curve can be identified with the *secant*, hence the formula:

$$(*) \quad (ds)^2 = (dx)^2 + (dy)^2.$$

Formula (*), for an *infinitesimal* line element ds is the starting point for a long chain of developments, with one key concept leading to the next. We'll ultimately see the descendents of formula (*) in their 20th century expression for a line element ds in 4-dimensional Minkowski space connected with the special theory of relativity:

(M)(for Minkowski) $\qquad ds^2 = dx_1^2 + dx_2^2 + dx_3^2 + dx_4^2,$

where $dx_4 = icdt, c =$ the speed of light in vacuum and $i^2 = -1$; and finally we'll encounter ds through an entirely novel concept, as part of the fundamental tensor of general relativity. This remarkable road of intertwining conceptual and technical innovations will take us through the key ideas of Euler, Gauss, Riemann, Ricci and Levi-Civita.

Let us go back to formula (*), or rather, to its 3-dimensional form:

(Em)(for Euclidean metric) $\qquad ds^2 = dx^2 + dy^2 + dz^2.$

In due time, formula (Em) was re-established on a rigorous basis in terms of derivatives or differentials, though arguing in terms of infinitesimals had and still has considerable heuristic and intuitive appeal. This is a standard trick of mathematicians, namely, to re-interpret expressions or formulas that were first obtained on a heuristic or less rigorous basis. Thus, if $x = x(t), y = y(t), z = z(t), (a \leq t \leq b)$, is the parametric representation of a curve C, where the three functions are assumed to admit continuous derivatives, then formula (Em) in full rigorous dress becomes

$$(ds/dt)^2 = (dx/dt)^2 + (dy/dt)^2 + (dz/dt)^2.$$

From ds/dt, the distance between two points on the curve C can be calculated by straightforward integration.

I'll continue to use the formula (Em) in the sequel, with the ds, dx, dy and dz now interpreted as *differentials*, as was first suggested by Cauchy (1789—1857), and has standardly been done ever since in the emergent field of *differential geometry*.[8]

"The exploration of physical problems", writes Kline [1972, p. 544], "led inevitably to the search for greater knowledge of curves and surfaces, because the paths of moving objects are curves and the objects themselves are

---

[8] The term "differential geometry" was introduced by Luigi Bianchi (1856–1928) in 1894.

three-dimensional bodies bounded by surfaces. The mathematicians, already enthusiastic about the method of coordinate geometry and the power of the calculus, approached geometric problems with these two major tools. The impressive results of the [18th] century were obtained in the already established area of coordinate geometry and [now, in] the new field created by applying the calculus to geometric problems, namely, differential geometry."

Some of the major initial developments in the study of space curves are due to the great Swiss mathematician Leonard Euler (1707–1783), essentially the father of differential geometry. In the manner of representing space curves by parametric equations, Euler also represented 3-dimensional *surfaces* in parametric form, by setting $x = x(u,v), y = y(u,v), z = z(u,v)$, with u and v running through a region in the (u,v)-plane. Such a parametric representation of surfaces is referred to as a *mapping* from a region in the 2-dimensional (u,v)-plane to the surface, as embedded in 3-dimensional space.

The word "mapping" is well chosen; as a matter of fact, the differential geometry of surfaces received its impetus from the actual, and practice-oriented study of map making and the determination of geodesics on surfaces, that is, finding the shortest arc between two points on a convex surface. Euler made significant advances in inventing the appropriate tools and concepts, in particular, for the study of *developable surfaces*, namely, surfaces that can be flattened out on a plane without distortions. But prior to the path breaking new orientation given by Gauss (1777–1855) to studies of differential geometry of surfaces, until then, surfaces were regarded just as boundaries of solids. Thus, Euler spoke of solids whose surfaces may be unfolded on a plane, and studied conditions permitting such mappings.

Now to Gauss.

4 Putting Pythagoras on surfaces

Starting with the publication in 1828 of the 80-page long memoir of Gauss entitled *Disquisitiones generales circa superficies curvas*[9] or in translation, "General investigation of curved surfaces", the *intrinsic geometry* of surfaces became the focus of a chain of new conceptual developments.

First some background information. Shortly after having published in 1801 his masterpiece *Disquisitiones arithmeticae* at the age of 20, opening

---
[9]For the original text of Gauss, with translation and commentaries, see Dombrowski [1979].

the modern era of algebraic number theory, Gauss was named director of the Göttingen astronomical observatory. In this capacity, his work required undertaking extensive land surveying, space measurements, and map making.[10] During this period, his early interest in the foundations of geometry turned to the question whether physical space was Euclidean or not.[11] The idea of an anti-Euclidean geometry, as it was called, took roots in Gauss's thinking, though he never dared to publish his ideas in this domain, but discussed it in correspondence with friends. However, Gauss's occupation with surveying gave him the impetus to develop his ideas on space in a non-controversial context, namely, the differential geometry of curved surfaces.

Let us turn now to the technical details of Gauss's 1828 "General investigation of curved surfaces". Consider a surface S given by parametric equations $x = x(u,v), y = y(u,v), z = z(u,v)$, and a curve C on the surface with parametric equations $u = u(t), v = v(t), (a \leq t \leq b)$.

By a straightforward calculation using the chain rule — well known at that time, one obtains:

$$\frac{dx}{dt} = \frac{\partial x}{\partial u}\frac{du}{dt} + \frac{\partial x}{\partial v}\frac{dv}{dt};$$

Similarly for dy/dt and dz/dt. Substituting these expressions in the analytic 3-dimensional Pythagorean relation

$$(ds/dt)^2 = (dx/dt)^2 + (dy/dt)^2 + (dz/dt)^2$$

and collecting terms, one obtains that

$$(ds/dt)^2 = E(du/dt)^2 + 2F(du/dt)(dv/dt) + G(dv/dt)^2,$$

where E, F, and G are functions of the various partial derivatives resulting from the chain rule.

Using the notation of differentials, as Gauss did, one obtains for ds an expression as a *quadratic differential form* as follows:

---

[10] "Once again", writes Jammer [1960, p. 150], "we see that historically viewed, abstract theories of space owe their existence to the practice of geodesic work, just as ancient geometry originated in the practical needs of land surveying."

[11] Indeed, Gauss first approached the question experimentally and undertook measurements of the triangle formed by the sides of three high mountains, in order to see whether they sum up to 180° or not. The experiment was inconclusive. For details, see Jammer [1960, p. 145] and Kline [1972, pp. 872–873].

(G) (for Gauss) $$ds^2 = Edu^2 + 2Fdudv + Gdv^2.$$

The relation (G) is now known under the name of the *First Gaussian Fundamental Form*.

Let us compare (G) with the analytic Euclidean-Pythagorean metric relation for a line element ds in Euclidean 3-space:

(Em) $$ds^2 = dx^2 + dy^2 + dz^2.$$

Notice that in (Em), (x, y, z) refer to rectangular coordinates in 3-space, whereas in (G), (u, v) are *local* curvilinear coordinates *on* the surface S. Now let us forget about how the functions E, F, and G were calculated in terms of partial derivatives, and consider them to be *independent* parameters that determine the geometry of the surface S. This was exactly the key idea of Gauss in his 1828 "General investigations of curved surfaces" memoir. In passing from the x, y, z three-dimensional Euclidean rectangular coordinate system to the u, v system in parametrizing surfaces, Gauss has liberated, so to speak, the surface from being embedded in Euclidean space. The surface is now a geometric object in itself, as schematized in Fig. 6 below, and the u, v parameters become local coordinates as part of the surface itself.

In other words, no more *external* rectangular coordinates as in (Em), but only *internal* coordinates as in (G). The inner, *intrinsic* differential geometry of a surface is studied now via expressions depending solely on the choice of the functions E, F, and G which determine the metric as given by the quadratic differential form in (G). Think of a bug living on a surface S with no surrounding external space in which the surface S could have been embedded. Our bug lives on and can only know "surface-land", by analogy to the famous "flat-land" of Edwin Abbott.

Guided by the already well-understood non-Euclidean geometry of the surface of a sphere, Gauss used spherical geometry as a *heuristic guide* and technical tool in order to define and study, such as, various notions of curvature of a surface, angles between "lines", invariants under a change of local coordinates, and so on.

In the words of Gauss himself: "... it is easily seen that the consideration of figures constructed upon the surface, their angles, their areas and their integral curvatures, their joining of points by means of shortest lines, and the

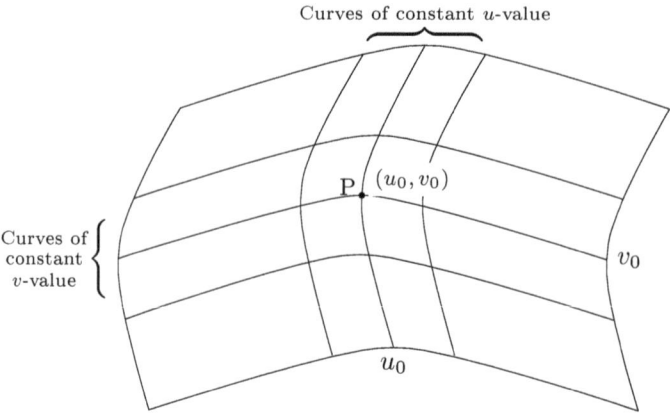

Figure 6. Gaussian coordinates on an arbitrary curved two-dimensional surface. Every point is located at the intersection of a unique u-curve and a unique v-curve, which determine the point's coordinates. (After Sklar [1974])

like, all belong to this case", [i.e., the relations "that are independent of the various forms which the surface may assume"]. "All such investigations must start from this, that the very nature of the curved surface is given by means of expressions of a linear element in the form $[Edp^2+2Fdpdq+Gdq^2]^{1/2}$."[12]

I won't discuss the various technical theorems proven by Gauss, but in closing the discussion here of his 1828 memoir, I'll just quote Morris Kline's succinct summary of the key conceptual points:

> "...given...u and v [as] coordinates on the surface and the expressions for ds^2 in terms of E, F, and G as functions of u and v, all the properties of the surface follow from this expression... What kind of geometry does the surface possess if it is regarded as a space in itself? *If one takes the 'straight lines' on the surface to be the geodesics, then the geometry is non-Euclidean...* Thus if the surface on the sphere is studied as a space in itself, it has its own geometry and even if the familiar latitude and longitude are used as coordinates of points, the

---

[12]Cited from the translation in Dombrowski [1979, p.89 & p.91].

geometry of that surface is non-Euclidean because the 'straight lines' or geodesics are arcs of great circles on the surface... What Gauss's work implied is that there are non-Euclidean geometries at least on surfaces regarded as spaces in themselves."
(Op. cit., p.888, emphasis added).

And now to the crux of the matter:

"If one can pick different sets of E, F, and G and thereby determine different geometries on the same surface, why can't one pick different distance functions in our own ordinary three-dimensional space? The common distance function in rectangular coordinates is, of course, $ds^2 = dx^2 + dy^2 + dz^2$, and this is obligatory if one starts with Euclidean geometry because it is just the analytic statement of the Pythagorean theorem. However, given the same rectangular Cartesian coordinates for the points of space, one might pick a different expression for ds^2 and obtain a quite different geometry for that space, a non-Euclidean geometry. The extension to any space of the ideas of Gauss first obtained by studying surfaces was taken up and developed by Riemann."
(Op. cit., p. 889).

5 Shrinking Pythagoras to the infinitely small

Bernard Riemann (1826–1866) was one of the greatest mathematicians of all times. The depth of his ideas, the technical prowess that he deployed in his numerous investigations, the entire new paths that he traced in the span of his relatively short life, all are absolutely astounding.

Here I am just concerned with the spectrum of problems which Riemann took over from Gauss, his mentor, concerning the nature of space, both abstractly and relative to the question whether physical space was Euclidean or not. Riemann presented his ideas on these issues in the habilitation lecture of 1854, entitled *Über die Hypothesen, welche der Geometrie zu Grunde liegen,* or in translation, "On the hypotheses which lie at the foundation of geometry".[13] The lecture was delivered to the entire philosophy faculty of

---

[13] See B. Riemann [1854]. English translation: "On the hypotheses which lie at the foundations of geometry". In: William K. Clifford, Mathematical Papers, pp. 55–69; Chelsea Publishing Co., 1968, Bronx, NY.

the University of Göttingen, in the presence of Gauss, just a year before the master's death. The lecture minimized mathematical technicalities that were subsequently elaborated by Riemann. In the opening section of the lecture, Riemann says: "I have in the first place... set myself the task of constructing the notion of a multiply extended magnitude out of the general notions of magnitude. It will follow from this that a multiply extended magnitude is capable of different measure-relations, and consequently that space is only a particular case of a triply extended magnitude. But hence flows as a necessary consequence that the propositions of geometry cannot be derived from general notions of magnitude, but that the properties which distinguish space from other conceivable triply extended magnitudes are only to be deduced from experience."

Riemann's way of expressing himself calls for a gloss. By "magnitude" *tout court* is meant here the ordinary real numbers, and "multiply extended magnitude" is Riemann's circumlocutory expression for n-tuples of real numbers. It was a complete novelty at that time, since in his abstract approach, Riemann wished to start right off with an n-dimensional space, of which physical 3-dimensional space with its Euclidean metric is just a special case. It took Riemann several pages to explain and justify the generalities of his innovation. Opting for the purely analytic approach, Riemann sets out to investigate first the general conceivable properties of an n-dimensional space with measures determined by a metric, and then leaves the actual decision whether physical space is Euclidean to empirical investigation. Thus, it was one of Riemann's objectives to show that Euclid's particular axioms were empirical rather than, as had been believed, self-evident truths.

In terms of notations subsequently used by Riemann in later publications, the starting point here is what Riemann has called an n-dimensional *manifold*, consisting of points $(x_1, x_2, \ldots x_n)$, and a metric for a line element ds given by the general quadratic differential form with

$$(R)\text{(for Riemann)} \qquad ds^2 = \sum_{i=1}^{n} \sum_{j=1}^{n} g_{ij} dx_i dx_j,$$

where $g_{ij} = g_{ji}$, but otherwise are *arbitrary* functions of the coordinates $(x_1, x_2, \ldots x_n)$ at a point, subject only to the obvious restriction that all factors on the right-hand side of (R) be positive. If $g_{ij} = 0$ for $i \neq j$ and all $g_{ii} = 1$, the relation (R) reduces to the ordinary Euclidean Pythagorean metric.

The geometry that Riemann developed, with the relation (R) as its point of departure, is now known as Riemannian geometry, or elliptic geometry. It is still a topic of actual research and further developments.[14] Riemann made use of concepts and techniques of Euclidean geometry by postulating that the geometric continuum under investigation should be approximately Euclidean in the neighborhood of each of its points. Thus, Riemann assumed the validity of the Pythagorean theorem "in the infinitely small". To state these ideas in a technically precise way required conceptual tools that were not yet on the tool shelves in the 19th century. But the path from heuristic developments to rigorous reformulations is only a matter of time, and then, moreover, rigor itself is time-dependent, and never absolute. Only with the development of general topology in the 20th century was it possible to reformulate appropriately the concept of an n-dimensional Riemannian differential manifold and recast Riemann's theory in a rigorous mold.[15]

Let us return now to the actual content of Riemann's 1854 habilitation paper. On the basis of the metric as defined by the relation (R), Riemann developed formulas for arc length of curves on a manifold, angles between curves, and most importantly, a formula for the curvature at a given point. All these quantities depend, of course, on the functions g_{ij}. In other words, different choices of the g_{ij} determine different geometries.

I won't state these various formulas here; they can be found, for instance, in Chapter VIII of Stoker [1969]. An important consequence of the formula for the curvature of a manifold is the following. For a Riemannian manifold of *constant curvature*, three cases are possible: (a) The curvature is zero, and the space is "flat" (in Riemann's terminology), i.e., the space is Euclidean, hence Euclid's axiom of parallels holds. (b) The curvature is a positive constant, and the space is spherical, or, as it was subsequently called, the geometry is elliptic. As on a sphere, where all great circles representing "straight lines" intersect, there are no parallel lines in elliptic geometry. (c) The curvature is a negative constant, and the space is hyperbolic. In such a space, there are infinitely many "straight lines" through a given point and parallel to a line not containing that point.

Equivalently, the tripartite classification of metric geometries of constant curvature can also be stated in terms of the sum of the interior angles on a triangle. If the sum is 180°, the geometry is Euclidean; if the sum

---
[14] See for instance the book by Klingenberg [1995].
[15] For details, see Stoker [1969].

exceeds 180°, the geometry is elliptic; in the remaining case, if the sum is less than 180°, the geometry is hyperbolic. These characterizations open the road for an empirical determination of the actual geometry of physical space. Such measurements were already undertaken by Gauss but remained inconclusive. (See note no. 11).

Riemann dealt at length with the important question concerning the *invariance* of the line element ds in the formula (R) under a change of coordinates. Obviously, the geometric properties of space ought to be independent of the way the space is described by the choice of a coordinate system. And similarly, in terms of applications— a key issue in the theory of relativity —physical laws ought to be independent of the position of the observer. Technically, this amounts to the following. Suppose that each point P on the manifold with coordinates $(x_1, x_2, \ldots, x_n)$ is given new coordinates $(y_1, y_2, \ldots, y_n)$ in a different coordinate system. The old coordinates are related then to the new coordinates by a system of equations that can be written as

$$(T) \qquad x_i = x_i(y_1, y_2, \ldots, y_n), \qquad i = 1, 2, \ldots, n.$$

Under this change of coordinates—assumed one-to-one—the functions g_{ij} depending in the x_i are transformed into functions h_{ij} depending on the y_i, and in the new coordinate system, the line element ds' is given by the formula

$$(R') \qquad ds'^2 = \sum_{i=1}^{n} \sum_{i=1}^{n} h_{ij} dy_i dy_j.$$

The crucial technical question which Riemann investigated is as follows. What conditions must be imposed on the transformation functions in (T) so that in relation to the ds of (R),

$$ds = ds',$$

in order to yield the same metric geometry, though its description via a choice of coordinates has changed? Even more difficult is the question of the conditions that ensure that one obtains, under a change of coordinates, the same value for the curvature and other geometric parameters. Riemann obtained necessary conditions, having invented new notational devices of a fairly complicated nature. Though he himself succeeded just in the initial steps, the question of the *invariance* of differential forms has now moved center stage.[16]

6 The workshop enriched: the new tools invented by Christoffel, Ricci and Levi-Civita

The line of the successors of Riemann in investigating invariance questions of differential forms starts with Eugenio Beltrami (1835–1900), Rudolf Lipschitz (1832–1903) and Elwin Bruno Christoffel (1829–1900). The latter set out to find necessary and sufficient conditions that permit transforming differential forms that include even higher differentials. To this end, he introduced a system of notations whose use simplifies the calculations and investigations concerning the invariance of forms. They are now known under the name of Christoffel symbols and have continued to play an important role in all subsequent studies of *differential invariants* in the most general setting. It soon became clear, however, that a general theory was called for. Some initial steps were already taken by Gregorio Ricci-Curbastro (1853–1925) in a series of publications as of 1885; and then, in a joint major work with his former pupil Tullio Levi-Civita (1873–1941), an extensive theory was presented in their lengthy memoir entitled "Méthodes de calcul différentiel absolu et leurs applications". (Ricci-Curbastro and Levi-Civita 1901). The Ricci/Levi-Civita absolute differential calculus has subsequently been renamed *tensor analysis,* the word 'tensor' coming from applications to elasticity. Essentially, a tensor is a set of functions or components, fixed relative to one frame of reference, that transform under change of coordinates in accordance of certain laws. One of the most important properties of tensors is the following. *If a tensor equation is true for one coordinate system, it is true for all coordinate systems.* With the aid of the tensor concept, one can re-express many concepts of Riemannian geometry in a simplified form, notably the formula for the curvature of space. I'll forgo the technical details here, but wish only to give a famous yet simple example of an invariance under a certain change of coordinates.

---

[16] For a more elaborated discussion related to Gauss's theory of surfaces and Riemann's theory, see the excellent exposition of Sklar [1974, Chapt. II]. Section B contains also many helpful diagrams and other material related to the rise of non-Euclidean geometries. The book of Sklar, as a whole, addresses readers with a philosophical bent and keeps mathematical technicalities to a minimum.

Consider again the metric of a line element in 4-dimensional Minkowski space of constant curvature, given by the equation

$$(M) \qquad ds^2 = dx^2 + dy^2 + dz^2 - c^2 dt^2,$$

where c is the speed of light in vacuum.

Under the Lorentz transformation, with

$$x = \frac{x' - vt'}{\sqrt{1 - (v/c)^2}}, y = y', z = z', t = \frac{t' - (v/c)^2 x'}{\sqrt{1 - (v/c)^2}}$$

a straightforward calculation shows that with this change of coordinates,

$$dx^2 + dy^2 + dz^2 - c^2 dt^2 = (dx')^2 + (dy')^2 + (dz')^2 - c^2 (dt')^2.$$

Thus, $ds = ds'$, i.e., (M) is *invariant* under the Lorentz transformation of special relativity.

One does not have to be an Einstein to perform this simple calculation; but, of course, having thought deeply about the significance of the Lorentz transformation from a physical point of view and the consequence of the invariance of the Minkowski metric, well, that's another story!

7 "A tool, a tool, my general relativity for a tool!"...

Thus Einstein could have exclaimed in early 1912. Here are the details of this fascinating case history.

The path, from special relativity of 1905 to the general theory of relativity of 1916 that incorporated into its fundamental equations expressions for the presence of gravitation, was an arduous and tortuous one, both with theoretical hurdles, false turns, and equally, beset by mathematical difficulties. When at last it became clear what ought to be done, Einstein wrote to Sommerfeld later in 1912 the following:

> "At present I occupy myself exclusively with the problem of gravitation and now I believe that I shall master all difficulties with the help of a friendly mathematician here [in Zurich]. But one thing is certain, in all my life I have labored not nearly as hard, and I have become imbued with great respect for mathematics, the subtler part of which I had in my simple-mindedness

regarded as pure luxury until now. Compared with this problem, [that is, the problem of gravitation], the original relativity is child's play".[17]

The "friendly mathematician" to whom Einstein alluded to was his lifetime friend, the Swiss mathematician Marcel Grossman (1878–1936), to whom Einstein had dedicated his doctoral dissertation. The friendship between Einstein and Grossman dated back to their student years at Zurich; it subsequently developed into an important technical cooperation, as related by Pais (1982).

In a later recollection in 1923, Einstein wrote: "I had the decisive idea of the analogy between the mathematical problem of the [general] theory of relativity and the Gaussian theory of surfaces only in 1912, however, after my return [from Prague] to Zurich, *without being aware at the time of the work of Riemann, Ricci and Levi-Civita*. This [work] was first brought to my attention by my friend Grossman". (Quoted from Pais [1982, p. 212]; emphasis added). Grossman taught Einstein the fundamentals of Riemannian geometry and the Ricci/Levi-Civita theory, the latter having been subsequently baptized by Einstein as tensor analysis.

In the language of tensor analysis, the fundamental line element in the general theory of relativity is written in the form

$$(E-G)(\text{for Einstein} - \text{Grossman}) \quad ds^2 = g_{\mu\nu} dx_\mu dx_\nu.$$

The quantities $g_{\mu\nu}$—where μ, ν range over all pairs from 1 to 4 and the summation sign is omitted by a convention introduced by Einstein—represent the dynamic fields related to gravitation. Einstein refers to these quantities $g_{\mu\nu}$ as the covariant fundamental tensor. Notice that $(E-G)$ is a tensor equation! Thus, *from the point of view of the general theory of relativity, the geometry of space is determined by the presence of matter,* and there followed from this the famous prediction by Einstein that light is bent in the vicinity of a strong gravitational field.

Let us take stock. We have followed the long road from the Pythagorean-Euclidean quadratic distance formula, through the fundamental Gaussian quadratic differential form in which the functions g_{ij} appeared first as E, F, and G, depending on the various partial derivatives; and then, via all the

---

[17]Cited by Pais [1982, p. 216].

successive metamorphoses in contexts and meanings of these functions and the line element ds, to arrive at their place in a tensor equation of Einstein's general theory of relativity.

There are these lucky moments when an experimentalist in need of certain material tools, or when a theoretician needs the appropriate mathematical machinery finds that the tools are already on the tools shelf. Be they material or thought tools, the analogy merits reflection. (See Marquis [1997] and [1999]).

To sum up:

Mathematicians manufacture conceptual tools, whatever be the motivation, such as "art for art's sake" or for solving practical problems. The dynamics is the same. New tools serve to solve existing problems that engender in turn new problems, calling for the development of further tools. And thus the spiral soars to further heights.

This is the path of the proliferous growth of mathematical knowledge.

BIBLIOGRAPHY

[Cellucci, 2000] C. Cellucci. The growth of mathematical knowledge: an open world view. In *The Growth of Mathematical Knowledge*, E. Grosholz and H. Breger (eds.), Dordrecht, Kluwer, 153–176, 2000.

[Dombrowski, 1979] P. Dombrowski (ed). 150 Years after Gauss' *disquisitiones generales circa superficies curvas"*. Astérisque 62. Paris, Société Mathématique de France, 1979.

[Dove, 2002] I. Dove. Can pictures prove? *Logique et Analyse*, **179–180**, 309–340, 2002.

[Gillies, 2000] D. Gillies. An empiricist philosophy of mathematics and its implications for the history of mathematics. In: *The growth of mathematical knowledge*, E. Grosholz and H. Breger (eds.), Dordrecht, Kluwer, 41–57, 2000.

[Jammer, 1960] M. Jammer. *Concepts of space*. New York, Harper & Brothers, 1960.

[Kline, 1972] M. Kline. *Mathematical thought from ancient to modern times*. New York/Oxford, Oxford University Press, 1972.

[Klingenberg, 1995] W.P.A. Klingenberg. *Differential geometry*. Berlin, Walter de Gruyter, 1995.

[Marquis, 1997] J.-P. Marquis. Abstract mathematical tools and machines for mathematics. *Philosophia Mathematica (III)*, **5**(3), 250–272, 1997.

[Marquis, 1999] J.-P. Marquis. Mathematical engineering and mathematical change. *International Studies in the Philosophy of Science*, **13**(3), 245–259, 1999.

[Michaels, 1978] A. Michaels. *Beweisverfahren in der vedischen Sakralgeometrie*. Wiebaden, Franz Steiner, 1978.

[Müller, 1930] C. Müller. Die Mathematik der Sulvasûtra: Eine Studie zur Geschichte indischer Mathematik. *Abhandlungen aus dem mathematischen Seminar der Hamburgischen Universität*, **7**, 173–204, 1930.

[Needham, 1959] J. Needham. *Science and Civilisation in China*. Cambridge, Cambridge University Press, 1959.

[Pais, 1982] A. Pais. The Einstein-Grossman collaboration. In: *'Subtle is the Lord...'* : *The science and life of Albert Einstein*, by A. Pais. Oxford, Oxford University Press: 208–227, 1982.

[Ricci-Curbastro, 1901] G. Ricci-Curbastro and T. Levi-Civita (1901). Méthodes de calcul différentiel absolu et leurs applications. *Mathematische Annalen*, **54**, 125–201, 1901.

[Riemann, 1854] B. Riemann. Über die Hypothesen, welche der Geometrie zu Grunde liegen. *Gesammelte mathematische Werke (Collected Works)*, 272–287, 1854.

[Sklar, 1974] L. Sklar. *Space, Time, and Spacetime*. Berkeley, University of California Press, 1974.

[Stoker, 1969] J.J. Stoker. *Differential Geometry*. New York, John Wiley & Sons, 1969.

Categorification as a Heuristic Device
DAVID CORFIELD

I was drawn into the philosophy of mathematics by the writings of Lakatos, Brunschvicg and Lautman. When I first started to read the mainstream English-language philosophy of mathematics literature, I was immediately struck by its almost complete lack of interest in what I considered to be the treasure house of mathematics. The philosophers I read seemed to think that the best way to get a handle on mathematics was to find some formal calculus or other which could be said to represent it in its totality. Set theory or second-order logic appeared to be just the thing for the job. To me this was like badly packing some pieces of jewellery in a huge cardboard box, then speaking indiscriminately of the whole of the space enclosed within the box as though it were the precious contents. It seemed to me that were we to pose the right questions, we would be required to look at the finer filigree. Or to put it the other way around, if we find we are not required to look at the details of mathematical thinking, we are not posing the right questions.

Michael Polanyi wrote in *Personal Knowledge*

> Between the practice of hackneyed exercises on the one hand and the heuristic visions of the lonely discoverer on the other, lies the major domain of established mathematics on which the mathematician consciously dwells by losing himself in the contemplation of its greatness. A true understanding of science and mathematics includes the capacity for a contemplative experience of them, and the teaching of these sciences must aim at imparting this capacity to the pupil. [Polanyi, 1958, pp. 195–196]

I have always been fascinated by the interface between these 'heuristic visions' and 'established mathematics'. Visions tell stories of how mathematics has come to look as it does and of how it ought to develop in the future. But, how does a novel vision get taken up by the rest of the community? In what sense can this taking up process be called rational? Those who dwell on the static appraisal of the truth of mathematics within an overarching framework are inclined to see these questions as lying wholly within the context of discovery, and so consider them outside the bounds of philosophy. I reject this charge completely. I am very much interested in justification, but not primarily at the level of single propositions where so much work has already been done. Rather, I want to explore how powerful visions can spawn thriving research programmes, in which fruitful concepts and theories can be developed. There can be no absolute standpoint from which judgements about these good qualities can be made. They take place in 'real time', and so one must enter into the thickets of the history of mathematical practice.

Those who agree with me can choose to study mathematical reasoning at a very local level, local in historical terms and in terms of the patch of mathematics studied. The risk here, however, is in losing oneself in the minutiae of specific fragments of reasoning. On the other hand, the temptation to over-generalise from a detailed case must be resisted. We need detailed historical studies of allusive examples of reasoning which represent the established and emerging styles of thinking of an epoch, along with accounts of the way they transmogrify (cf. Kvasz's paper in this volume).

I also recognised that it was very important that we do not neglect the relatively recent past. Otherwise, the impression that mathematics has stabilised and can be treated now in timeless fashion would not go properly challenged. In recent decades, with the proliferation of mathematics, the issue that has come to the fore has been how best to organise the field. What I sought, then, was a reasonably large-scale contemporary effort to organise the contents of mathematics. One that makes genuine contact with, and goes along the grain of, valued findings. One that gives a retrospective account of the development of episodes from the past, and that suggests lines for future expansion. Those who have read my book, *Towards a Philosophy of Real Mathematics*, will know that I think I've found the answer in 'higher-dimensional algebra'.[1]

Now, I certainly do not wish to claim that higher-dimensional algebra has all the answers to the way mathematics is today, how it has emerged and the path it is travelling along, but I do believe that it possesses sufficiently many noteworthy attributes that a number of philosophers might usefully

---

[1] For the visionary aspect of higher-dimensional algebra, see in particular [Corfield, 2003, p. 249], and also its elaboration in section 3 of [Corfield, forthcoming].

engage with it. If we could have just half the resources currently allocated to the reconstruction of mathematics in terms of second-order logic, we would be able to cast in sharp relief a host of important features. These features, in my opinion, merit the kind of attention Russell lavished on the logics of Cantor, Frege and Peano.

In chapter 10 of my book, I began to sketch some of these features. More recently, when invited to speak to a group of sixty or so mathematicians in Minneapolis, who had gathered together to compare the dozen or so existing definitions of higher-dimensional categories (see http://www.ima.umn.edu/categories), I outlined two broad approaches as to how mathematics might come to inspire philosophy once more. One of these approaches would attempt to link mathematics via philosophy of science to epistemology, aesthetics and ethics; the other, I dubbed the *Russellian* approach, would use higher-dimensional algebra to run through traditional trading links between mathematical logic and philosophy, reforming metaphysics on the way. Here are suggestions for the Russellian approach:

(a) 'Every interesting equation is a lie': behind every interesting equation there lies a richer story of isomorphism or equivalence. Much more subtle conceptions of sameness are made available by higher-dimensional algebra.

(b) Reappraisal of property and structure: maps between different categories of entity factor to reveal what is property and what structure. This is important to understand the difference between abstraction and generalisation.

(c) New insights into logic: for example, a form of modal logic appears in a natural sequence, coming after propositional logic and predicate logic. Just as toposes throw light on constructive logic, their categorifications, i.e., 2-toposes should be important.

(d) New notions of complex structures for biology, neurology, computer science, etc.: as yet perhaps the most speculative of the entries here.

(e) Part-whole relations, the nature of space: one of the inspirations for higher-dimensional algebra was the revolutionary vision of Alexandre Grothendieck.

(f) Conceptions of space are currently being radically transformed. Higher-dimensional algebra is finding increasing use in leading approaches to quantum gravity.

(g) Diagrammatics to question the 'transparency' of logic: the blurring of the distinction between topology and algebra in the notation of higher-dimensional algebra reveals its iconic and symbolic roles (see the paper by Emily Grosholz, this volume).

(h) 'The sphere spectrum is the true integers': this is an example of taking the process of categorification to its extreme, starting out with the integers. It illustrates the idea that higher-dimensional algebra can carve out the 'right' concepts. If you start out with an important construction and categorify it 'well', you will end up with something else important.

In the brief space of this paper I can hardly begin to indicate the full significance of any one of these features. Instead, I'll give an overview of some of them. As ever when I write about higher-dimensional algebra, I am enormously indebted to John Baez and James Dolan for their expository work. My borrowings are too numerous to detail. I strongly advise the interested reader to look up their papers and web material at Baez's home page http://math.ucr.edu/home/baez/.

Higher-dimensional algebra

In higher-dimensional algebra, also known as higher-dimensional category theory, you encounter a ladder which you're irresistibly drawn to ascend.[2] Let us begin with a finite set. About two elements of this set you can only say that they are the same or that they are different. Thinking about sets a little harder, you are led to consider what connects them, namely, functions (or perhaps relations). Taken together, sets and functions form a category. Now, there are two levels of entity, the objects (sets) and the arrows (functions), satisfying some conditions, existence of identity arrows and associative composition of compatible arrows.

There are plenty of examples of categories. For example, categories of structured sets, such as groups and homomorphisms, but also spatial ones,

---

[2] For a more leisurely account of this material, read Baez's 'Tale of n-categories', starting at http://math.ucr.edu/home/baez/week73.html .

such as 2-Cob, whose objects are sets of circles and whose arrows are (diffeomorphism) classes of surfaces between them. A pair of trousers represents an arrow from a single circle (the waist) to a pair of circles (the trouser cuffs) in this latter category.

In a (small) category, the collection of arrows between two objects A and B, $\mathrm{Hom}(A, B)$, forms a set. The only choice for two arrows in $\mathrm{Hom}(A, B)$ is whether they are the same or different. At the level of objects, however, there is a new option. A and B need not be the same, but there may be arrows between them which compose to the identity arrows on each object. This kind of sameness is often called isomorphism. So-called 'structuralists' are aware of this degree of sameness.

Let us continue up the ladder. Consider categories long enough, and before you know it you're thinking of functors between categories, and natural transformations between functors. Functors are ways of mapping one category to another. If the first category is a small diagram of arrows, a copy of that diagram within the second category would be a functor. Natural transformations are ways of mediating between two images within the target category. Taking all (small) categories, functors, and natural transformations, we have an entity with three levels, which we draw with dots, arrows and double-arrows. This is an example of a 2-category, the next rung of the ladder. In this setting, whether objects are the 'same' is not treated most generally as isomorphism, but as equivalence. Between equivalent objects there is a pair of arrows which do not necessarily compose to give identity arrows, but do give arrows for which there are invertible 2-arrows to these identity arrows.

This is something structuralists have missed. At the level of categories, isomorphism is too strong a notion of sameness. Anything you can do with, say, the category of finite sets, you can do with the equivalent (full) subcategory composed of a representative object for each finite number and functions between them.

Then, one more step up the ladder, 2-categories form a 3-category, with four levels of entity. Another example of a 3-category is the fundamental 3-groupoid of a space. Take the surface of a sphere, such as the world. Objects are points on the globe. 1-arrows between a pair of objects, say the North and South Poles, are paths. 2-arrows between pairs of 1-arrows, say the Greenwich Meridian and the International Data Line, are ways of sweeping from one path to the other. Finally, a 3-arrow between a pair of 2-arrows, say one that proceeds at a uniform rate between the Greenwich Meridian and the International Data Line and the other that tarries a while over New Delhi, is represented by a way of interpolating between these sweepings.

Mathematicians hate to stop a good thing when it's rolling, so are aiming to extend this process infinitely far to omega-categories, by defining them at one fell swoop. The idea for doing so was inspired by Alexandre Grothendieck, who realised that there was a way of treating spaces up to homotopy in algebraic terms. Already at the 2-category level there are many choices of shape to paste together. There are thus many ways of defining an omega-category. At present, twelve definitions have been proposed. It is felt, however, that the choice is in a sense immaterial, in that all ways will turn out to be the 'same' at the level of omega-categories, although each may be best suited to different applications. I was invited to a fortnight long workshop, hosted by the Institute for Mathematics and its Applications in Minneapolis, whose aim was to compare these definitions, and assess their potential applications. Regarding the latter, computer scientists from the French nuclear industry are using omega-categories to analyse potential deadlocks in multi-processor computations.

Adding in some structure

Let's look at some applications not too far up the ladder. Plenty of categories have extra structure. Sets do not just form a category, but a highly structured kind of one, known as a topos. Toposes are environments for constructive reasoning. At a less structured level, we can view the category of sets as having two obvious binary operations, disjoint union and cartesian product. If we just take one of these and analyse the way the binary operation coheres with the functions, we extract what is called a monoidal category. These possess a 'multiplication' acting on objects, such as the tensor product of Hilbert spaces or vector spaces. A way to generate a monoidal category is to take any object A in a 2-category and consider only the arrows from A to A, and the 2-arrows between these. Reindex these 1-arrows so that they count as objects, and the 2-arrows as 1-arrows and you have a monoidal category. This process of killing off the lowest level is known as *delooping*. It can be systematised to allow the generation of the following table. Delooping moves you in a south-westerly direction.

Table 1. k-tuply monoidal n-categories

	$n = 0$	$n = 1$	$n = 2$	...
$k = 0$	sets	categories	2-categories	
$k = 1$	monoids	monoidal categories	monoidal 2-categories	
$k = 2$	commutative monoids	braided monoidal categories	braided monoidal 2-categories	
$k = 3$	" "	symmetric monoidal categories	weakly involutory monoidal 2-categories	
$k = 4$	" "	" "	strongly involutory monoidal 2-categories	
$k = 5$	" "	" "	" "	
...				

[Baez and Dolan, 1999]

This table comes in three flavours. The first is as above. The second requires inverses at each level of arrow. But the most interesting of the three requires a weaker form of inverse known as a dual, a generalised form of the adjoint construction from Hilbert space theory. 2-Cob has these. For instance, the adjoint of a pair of trousers is another pair of trousers oriented in the opposite direction. Gluing these arrows at the trouser cuffs, you see a map from the waist of one pair to the waist of the other. This is not an identity arrow from the waist to the waist, nor is it equivalent to such an identity.

The mathematics relevant to the (k, n) position concerns the world of n-dimensional things living in an $(n + k)$-dimensional world. For example, in position $(1, 2)$ of the table flavoured with duals, we are dealing with lines and circles living in a 3-dimensional world. This is where knots, loops and tangles live. To find ways of distinguishing them, we need something algebraic which belongs to that same position. In the 1980s candidates were found, namely, representations of quantum groups. What is so unusual about this work is that the algebra has to be tailored to the dimension. Where ordinary algebraic topology was happy to use groups to pick up information in all dimensions, so-called *quantum topology* requires specific kinds of algebra for specific dimensions.

The pay-off is two way. The quantum groups help classify knots, while tangles help us calculate with quantum groups. Regarding the latter we are increasingly finding dimensioned notation appearing in textbooks, such as the following 'proof' by [Majid, 1995, pp. 444–446]

To write out the steps of this calculation in the usual linear form would be horrendous. We can think of the notation used in this process as both an iconic representation of the topological object, the tangle, while at the same time a symbolic representation of the algebra, a mapping between two algebraic entities. This is an excellent example of the phenomenon Emily

Grosholz treats in her paper. Higher-dimensional algebra also throws light on the iconic aspects of very simple symbolism, where the letters of algebra and logic are zero-dimensional entities living in some higher-dimensional environment (see [Corfield, 2003, pp. 242–251] for further details).

Categorification

Given that mathematicians and physicists want to know about how spheres can tie up in 4-dimensional space, they would be keen to discover a method of moving eastwards on the n-categories table. Such an operation is known as *categorification*. Categorified quantum groups would constitute the right algebra for the $(2, 2)$ position. The opposite process, known as *decategorification* is algorithmic. Take your n-category, throw away all top level arrows apart from invertible ones, then collapse the $(n-1)$-arrows into equivalence classes. E.g., take the category of finite sets, retain only bijections, then count as the same any objects (0-arrows) linked by bijections. You should be able to see that what emerges is the set of natural numbers. In the process there is a loss of information. Can we reverse this process? Can any construction be categorified? No, there is no recipe. In fact, there is always more than one answer, although at times a best one is all but forced upon you.

The job of categorification may not be at all easy. At present, people are working hard to categorify the tools of physics: groups, Lie groups, Lie algebras, vector bundles, Hilbert spaces, etc. It is intriguing to find proponents of the two leading theories of quantum gravity, loop quantum gravity and string theory, discussing the role of 2-groups in higher-dimensional gauge theory,[3] but ordinary gauge theory is beyond most people's ken, so let us here run through some simpler examples:

(a) **Algebraic topology from mid-19th century to 1930** This is an example of how one may reconstruct episodes from the past. Another categorification of the natural numbers besides the category of finite sets is the category of finite-dimensional vector spaces over a given field. Riemann had characterised the torus as the closed orientable surface which could be made simply connected by cutting it along two lines to produce a rectangle. But associating '2' to the torus tells us nothing about the maps between two tori. By 1930, mathematicians

---

[3]This can be read in the archives of the sci.physics.research newsgroup, and also 'The String Coffee Table' blog at http://golem.ph.utexas.edu/string/.

were now associating the two-dimensional real vector space R^2 to the torus. Corresponding to any map between tori is a certain linear map between vector spaces, i.e., one representable as a 2 by 2 matrix. This association can be used to distinguish between mappings. You can also use these ideas to show that there is no homeomorphism between the torus and the sphere, since there can be no linear mapping between R^2 and R^2 which factors through R^0.

(b) **From series to species** This example derives from the work of categorically-minded mathematicians. Consider series in one variable. Let's take those of the form $\Sigma a_i x^i/i!$, a_i a natural number. To categorify we need to replace the a_i by sets of that cardinality. What you do is define a *species* (first defined by André Joyal) as something which when fed a collection will give you back the set of structures of a certain kind that can put on that collection. For example, the species *singleton* spits out a one-element set if given a collection of one element, but otherwise gives the empty set. It is denoted X. The species *set* when given a collection spits out the set of ways of putting a set structure on this collection. There's only ever one way of doing this. All the resulting sets have cardinality 1, so *set* is a categorified version of the series with $a_i = 1$, in other words of the exponential function $\exp x$. A species which only does this for even-numbered sets is the categorification of $\cosh x$, and for odd-numbered sets $\sinh x$. On the other hand, the species of ordered sets or permutations spits out the set of $n!$ such structures when fed an n element set. This is a categorified version of $1/(1-x) = 1 + x + x^2 + \ldots$ Finally, the species *cycle* which gives the set of ways of ordering a collection in a circle, where there is no distinguished first element, is a categorification of the series $x + x^2/2 + x^3/3 + \ldots$

Species compose in different ways:

$F + G$ places an F-structure *or* a G-structure on a collection. E.g., $2X = X + X$ can be thought of as colouring a singleton in two colours, say, either a blue singleton or a red singleton.

$F \cdot G$ places an F-structure on part of the input and a G-structure on the remainder. $\cosh^2 X$, for example, forms ordered pairs of even-numbered sets out of the input.

$F \circ G$ collects the ways of partitioning the input, placing a G-structure on each component of the partition, and placing an F-structure on the collection of components. For example, $\text{Exp}(2X)$ when given a collection looks to form a set of 2-coloured singletons.

With these tools we can now categorify some identities:

$\cosh 2x = \cosh^2 x + \sinh^2 x$ is a decategorification of the fact that the set of ways of 2-colouring an even-numbered collection is naturally isomorphic to the union of the set of ways of partitioning the collection into two even-numbered sets and the set of ways of partitioning the collection into two odd-numbered sets.

The following calculation is a decategorification of the fact that the species of permutations on a collection is isomorphic to the species which gives the ways of writing the elements of the collection in the form of a set of cycles:

$$1+x+x^2+\ldots = 1/(1-x) = exp(-ln(1-x)) = exp(x+x^2/2+x^3/3+\ldots)$$

We already know that any permutation can be written as a set of disjoint cycles, but the species isomorphism gives us extra information. Just because we have an identity of series does not mean it is a decategorification of an isomorphism of species on collections. For example, $\cosh^2 x = \sinh^2 x + 1$, tells us that for a non-empty collection, there are as many ways of splitting it into two even-numbered sets as there are ways of splitting it into two odd-numbered sets. But we require more structure on our input before we can talk about an isomorphism.

In a paper [Corfield, forthcoming] I have written on the subject of 'natural kinds' in mathematics, I have proposed that we draw a distinction between 'law-like' and 'happenstantial'[4] mathematical facts. We could use higher-dimensional algebra to make claims such as:

> *A happenstantial equation is one which cannot be categorified productively.*

The fourth triangular number, 10, is one more than the third square number, 9. Were there a law-like relation occurring here, something like the $(n+1)$th triangular number is one greater than the nth square

---

[4] My thanks to Jeremy Butterfield for suggesting the term.

number, you would be very confident that there would be a systematic species isomorphism. On the other hand, one might observe that the number of ways of 2-colouring a 6 element set is 64, while the number of ways of partitioning a 6 element set into two sets each with an even number of elements is 32, and there are also 32 ways to partition it into two sets with an odd number of elements. $64 = 32 + 32$. Is this just happenstantial? No, there is a species isomorphism lurking behind the scenes.

A further illustration of this phenomenon of categorification and law-likeness begins with the observation that the third coefficient of the j-function from nineteenth century work on the moduli space of elliptic curves is one greater than the dimension of the smallest non-trivial irreducible representation of the monster finite simple group (see [Corfield, 2003, pp. 125–126]). Is this happenstantial? No, a regularity was found linking further coefficients with the dimensions of the irreducible representations of the monster, which is now understood to be a decategorification of an isomorphism at the level of modules.

Species are having a large impact on the field of combinatorics. Baez and Dolan have taken the next step by categorifying species to *stuff types*, which they use to analyse the combinatorics of Feynman diagrams.

(c) **Categorifying logic** Looking at the n-categories table, you may wish for aesthetic reasons to fill in a missing triangle at the top left hand corner. Let's try to extend the top line to the left for the table taken with inverses. Here, instead of 'category' we should write 'groupoid'. As Emily Grosholz has described in her [1985], there are strong correspondences between logic and topology. Let's use topology to find what should be the entry to the left of 'Set'.

When I talked earlier about the ways of constructing a 3-category out of paths, and paths between paths, on the surface of the Earth, I was producing what is known as the fundamental 3-groupoid of the sphere. Let's go down a couple of dimensions. Take a space X and form the groupoid whose objects are the points of X, and whose arrows are classes of paths between two points, where any two paths which can be deformed to each other within X keeping the end points fixed are taken to belong to the same path. On the Earth, the Greenwich

Meridian and the Date Line would be counted as the same path. This *fundamental groupoid* is used to measure whether there are any ways of mapping a circle into X which cannot be extended to the mapping of a complete disk. Down a dimension, we look for the set of connected components of X. This is counting as the same any two points which form the end points of a line in X. In other words, it measures whether mappings of the 0-sphere (2 points) can be filled by the 1-disk (a line). To descend a further dimension we need to complete the last of the following sentences:

$\Pi_1(X)$ is a groupoid, which measures ways that circles S^1 can't be filled by D^2.
$\Pi_0(X)$ is a set, which measures ways that S^0 can't be filled by D^1.
$\Pi_{-1}(X)$ is a ?, which measures ways that S^{-1} can't be filled by D^0.

Well D^0 is just a point, and so its boundary S^{-1} is the empty set. $\Pi_{-1}(X)$ measures whether when the empty set maps into X it can be filled by a point. In other words it asks of X, are you empty? This is a yes/no question. The '?' must be filled with 'truth value'.

With this result in mind, and the idea that modal logic is a categorification of predicate logic, I worked out with John Baez one evening in Minnesota how it would have to look. There are more reasons as to why it looks right to categorify the way we did, but I hope the following table is suggestive.

Logic Extension	Propositional	Predicate (typed)	Modal Predicate
Groupoids			$S, T, \ldots$
Sets		$A, B, \ldots$	$A_S(u), B_{T,U}(v, w)$
Truth values	$P, Q, \ldots$	$P_A(x), Q_{A,B}(x, y)$	$P_{A(u)}, Q_{A(u),B(v,w)}(x, y)$
Axioms	$P \& Q \to R$	$\forall x \in A.\exists y \in B.Q(x, y)$	$\forall u \in S.\exists x \in A(u).P(x)$
Models	Set (lines of truth table)	groupoid	2-groupoid

The 'Axioms' line gives an example of an axiom of a theory expressible in the logic. The version of predicate logic is most naturally constructed as a typed logic. This accords with the thinking of theoretical computer scientists, who view philosophers as crazy to deal with some kind of universal set, relying on predicates to do the work of types. The highest level of syntax in the modal logic, such as S, refers to an accessibility relation between worlds. Notice there may be more than one of these. Researchers in artificial intelligence use such a logic to describe multi-agent systems. A predicate symbol is typed by an accessibility relation and has to be interpreted as a set at each world covered by the relation. I am not certain if this logic has ever seen the light of day before. It's something like a modal typed predicate S5n, but where the n equivalence relations jointly cover the set of worlds but need not do so individually.

This table suggests a highly degenerate logic to the left, with no symbols, and two models — true and false. It also suggests a kind of meta-modal logic to the right. If we worked with categories rather than groupoids, it seems as though we would arrive at a parallel S4 kind of modal logic, but details need to be checked. Another possibility is to work with directed homotopy, rather than ordinary homotopy.[5]

Can it be a coincidence that Charles Peirce's existential graphs grew in complexity from the alpha part to the beta part to the gamma part, which deal respectively with propositional logic, predicate logic and modal logic. Did he presage the first few steps of a ladder of categorifications of logic?

Conclusion

I certainly do not wish to give the impression that higher-dimensional algebra has a monopoly on the large-scale transformations possible in mathematics. Other candidates include *q-deformation*, finding non-commutative versions (e.g., C*-algebras are non-commutative topological spaces), and many discussed by Vladimir Arnold in his fascinating 'Polymathematics' paper [Arnold, 1999]. I do, however, think this is an extremely important one.

Some of the issues it raises do seem to touch on what might be called 'foundational' questions, if the term weren't reserved by others. But perhaps it is time to recover the term. Listen to the highly respected Russian mathematician Yuri Manin:

---

[5] For this suggestion and many other relevant ideas see [Porter, 2003].

> I will understand 'foundations' neither as the para-philosophical preoccupation with the nature, accessibility, and reliability of mathematical truth, nor as a set of normative prescriptions like those advocated by finitists or formalists. I will use this word in a loose sense as a general term for the historically variable conglomerate of rules and principles used to organize the already existing and always being created anew body of mathematical knowledge of the relevant epoch. At times, it becomes codified in the form of an authoritative mathematical text as exemplified by Euclid's Elements. In another epoch, it is better expressed by the nervous self-questioning about the meaning of infinitesimals or the precise relationship between real numbers and points of the Euclidean line, or else, the nature of algorithms. In all cases, foundations in this wide sense is something which is relevant to a working mathematician, which refers to some basic principles of his/her trade, but which does not constitute the essence of his/her work. [Manin, 2002, p. 6]

Manin continues by discussing the foundational role of Cantorian sets, before proceeding to explain how this role was taken up first by categories, and now by higher-dimensional categories.

Philosophers, we know so little about the mathematics of the past seventy years. Higher-dimensional algebra presents you with the opportunity to work your way quite rapidly to be able to deal with extremely important ideas. Even if you choose to look at other features of contemporary mathematics, it provides an excellent platform for your research. We need to understand how the 'vision thing' works in mathematics. By drawing philosophical attention to it, we can prompt existing and future mathematicians to be more open about their visions, or 'bottom lines' to use a term of Gian-Carlo Rota, which can only help them, us and our successors.

BIBLIOGRAPHY

[Arnold, 1999] V. Arnold. Polymathematics: is mathematics a single science or a set of arts? In *Mathematics: Frontiers and Perspectives*, V. Arnold, M. Atiyah, P. Lax and B. Mazur, eds. Pp. 403–416, American Math. Soc., 1999.

[Baez and Dolan, 1999] J. Baez and J. Dolan. Categorification. In *Higher Category Theory*, E. Getzler and M. Kapranov, eds. Pp. 1–36. Providence, RI: American Mathematical Society, 1999.

[Corfield, 2003] D. Corfield. *Towards a Philosophy of Real Mathematics*, Cambridge University Press, 2003.

[Corfield, forthcoming] D. Corfield. Mathematical Kinds, or Being Kind to Mathematics, available at http://philsci-archive.pitt.edu/view/year/2004.html , forthcoming.

[Grosholz, 1985] E. Grosholz. Two Episodes in the Unification of Logic and Topology, *British Journal for the Philosophy of Science*, **36**, 147–57, 1985.

[Majid, 1995] S. Majid. *Foundations of Quantum Group Theory*. Cambridge University Press, 1995.

[Manin, 2002] Y. Manin. *Georg Cantor and His Heritage*, 2002. http://arxiv.org/abs/math.AG/020924

[Polanyi, 1958] M. Polanyi. *Personal Knowledge. Towards a Post Critical Philosophy*. Routledge, 1958.

[Porter, 2003] T. Porter. Geometric Aspects of Multiagent Systems, *Electronic Notes in Theoretical Computer Science*, **81**, 2003.

Heuristics and Mathematical Discovery: The Case of Bayesian Networks

DONALD GILLIES

1 Introduction

I will begin this paper by discussing some ideas to be found in two recent books on the philosophy of mathematics. These are (i) Carlo Cellucci's *Filosofia e matematica*, published by Laterza in 2002, and (ii) David Corfield's *Towards a Philosophy of Real Mathematics*, published by Oxford University Press in 2003. I will start with Cellucci's book.

In his book, Cellucci is highly critical of the traditional or foundational approach to the philosophy of mathematics, based on the attempt to justify mathematics. Instead he advocates what he calls the heuristic approach to the philosophy of mathematics. As he says (2002, p. viii):

> According to the dominant point of view the principal problem in the philosophy of mathematics is that of the justification of mathematics. ... In this book I maintain instead that the principal problem of reflection on mathematics is that of mathematical discovery. This problem includes the problem of justification ...

I partly agree and partly disagree with this. It is certainly true that traditional philosophy of mathematics focussed exclusively on the problem of the justification of mathematics and neglected the problem of mathematical discovery. So I definitely think that philosophers of mathematics should now take up the problem of mathematical discovery and that interesting results are to be expected from investigating it. On the other hand, I do not think that the problem of discovery includes that of justification. So I hold that the problem of justification should remain on the agenda of philosophers of mathematics, as a problem partly related to, but partly separate from that of discovery. For the purpose of this paper, however, I want to emphasize my agreement with Cellucci and to adopt his heuristic approach.

Now the obvious objection to the claim that philosophers should study the problem of mathematical discovery is that discoveries in mathematics depend on psychological factors such as insights of genius, the subjective intuitions of creative mathematicians and so on; and that, consequently, mathematical discovery cannot be given a systematic philosophical treatment. Cellucci strongly challenges this point of view in the following passage [2002, p. xvii]:

> According to the dominant point of view mathematical discovery is an irrational process, which is not based on logic but rather on intuition. ...In this book I maintain instead that mathematics is a rational activity at every moment, including the most important, discovery. Since antiquity many have recognised not only that mathematical discovery is a rational process, but also that a method exists for it, namely the analytic method. This method gave a great heuristic power to the ancient mathematicians for the solution of geometrical problems, and has had a decisive role in the new developments of mathematics and physics at the beginning of the modern era. In it logic plays an essential role in the discovery of hypotheses, though this is not logic understood in the restricted fashion ... but in a wider fashion which includes also and above all non-deductive inferences.

Cellucci does not merely advocate a heuristic approach to the philosophy of mathematics, but actually makes a start with developing it, particularly in Chapters 30 to 38 of his book. Here he lists and illustrates quite a number of principles which he regards as fruitful for mathematical discovery. This investigation of Cellucci's does indeed call into question the claim that mathematical discovery is exclusively a matter of subjective intuitions and the like. There is however a point which can be regarded as doubtful. Cellucci makes clear in the passage just quoted that he believes that the principles underlying mathematical discovery are logical in character, so that there is, in effect, a logic of mathematical discovery. However, another point of view would be that there are indeed principles underlying mathematical discovery but that these principles are heuristics, or guides to discovery, which are not logical in character. It is not an easy matter to decide between these two points of view, since it is not clear what we should regard as constituting logic. If there is to be a logic of mathematical discovery, then logic will certainly, as Cellucci stresses, have to extended to include non-deductive inferences. Yet how far can we extend logic beyond its core of deductive inferences while still retaining something that is recognisably logic? Is there an inductive logic for example? And if so, what is its character? More generally what are the boundaries of logic? In the last section of this paper (Section 6) I will come back to this question and discuss some of the interesting ideas of Ladislav Kvasz on this subject. However for the moment, I will take the goal to be that of elucidating some of the heuristic principles involved in mathematical discovery, and leave aside the question of whether these principles should be regarded as logical in character.

Let me now turn to Corfield's new book. This contains a mass of interesting material ranging from automated theorem proving, through Bayesianism applied to mathematics, to a consideration of groupoids and higher-dimensional algebra. However, for the purposes of this present paper, I want to consider only one general methodological point which Corfield makes towards the beginning of his book. He points out that the mathematics considered by philosophers of mathematics tends to be almost exclusively the foundational mathematics of the period 1880–1930, and that, in particular, the mathematics of the last 70 years is largely ignored except perhaps, in some cases, for a consideration of further developments of foundationalist mathematics. As Corfield himself says [2003, p. 5]:

> By far the larger part of activity in what goes by the name *philosophy of mathematics* is dead to what mathematicians think and have thought, aside from an unbalanced interest in the 'foundational' ideas of the 1880–1930 period, ...

Corfield calls this attitude 'the foundationalist filter'. This filter removes from the attention of philosophers of mathematics any mathematics which is not foundationalist. Corfield thinks that philosophers of mathematics should remove this filter and consider mathematics which is not foundationalist. This could be some of the mathematics of the past, but Corfield recommends very strongly that philosophers of mathematics should take an interest in the non-foundationalist mathematics of the last seventy years which he thinks that they have hitherto largely ignored. As he says [2003, pp. 7–8]:

Straight away, from simple inductive considerations, it should strike us as implausible that mathematicians dealing with number, function and space have produced nothing of philosophical significance in the past seventy years in view of their record over the previous three centuries.

Corfield attempts in his book to redress the balance by considering from the philosophical point of view many developments in mathematics during the last seventy years.

That concludes my discussion of some of the ideas in the new books by Cellucci and Corfield. I will now explain how they have led to the plan for the present paper. Essentially I have taken from Cellucci the idea of studying the heuristics of mathematical discovery, and I will try to add to his treatment by considering an example of mathematical discovery different from the ones which he considers. Following the recommendations of Corfield, I have taken this example form the field of non-foundational mathematics in the last seventy years. The example in fact comes from my own favourite branch of mathematics: probability theory. Probability theory is usually considered by philosophers of science rather than philosophers of mathematics, and there are obvious reasons for this. Probability is closely connected to induction whose analysis, or in some cases denial, is a central issue in philosophy of science. Probabilities also appear in many scientific theories, notably quantum mechanics. But despite its interest for philosophers of science, probability theory is after all a branch of mathematics and an important one. So there may be some value in considering some of the general problems of the philosophy of mathematics in relation to probability theory.

Since I started studying probability theory in the 1960s, the most important development in the field has been, in my opinion, the discovery of Bayesian networks, which took place in the 1980s — fortunately well within the Corfield limit of seventy years. Many mathematical discoveries are of proofs of theorems, but some discoveries are of new mathematical concepts which give rise to new theories involving many theorems and having many uses in different areas. The most famous discovery of this type is perhaps the discovery of the group concept. The discovery of the concept of Bayesian network has this character. It has resulted in the development of an entirely new branch of probability theory which is now expounded in textbooks like Neapolitan 1990. None of the contents of Neapolitan 1990 would have appeared in a textbook of probability theory written before the 1980s. We have something here that is really new and that has also been applied with great success in a wide variety of different areas. We are thus dealing with a discovery of considerable importance and an analysis of the heuristics which led to this discovery may be not without some interest. In the next Section 2, I will give a brief historical account of how the discovery of Bayesian networks was made. This should also serve as an introduction to the concept for those who have not met it as yet. Then in Sections 3,4 and 5, I will state and analyse three heuristics which seem to me to have been involved in the discovery.

2 The Development of Artificial Intelligence and the Discovery of Bayesian Networks

One route which led to the discovery of Bayesian networks began with investigations into artificial intelligence (AI). This is the route which I will describe in what follows. The full story however is more complicated. There was another largely independent route which began with investigations into decision theory and which led to concepts not dissimilar from Bayesian networks. Another strand in the story is constituted by attempts to find economical ways of storing probability distributions in computers. The developments which I will describe, however, were largely self-contained and are suitable for analysis from the point of view of the heuristics involved. I will therefore leave the full account as the task for a more detailed history.

Research in AI began in the 1950s and many important ideas were developed by the pioneers. Then in the 1970s a breakthrough was produced by the creation of expert systems. The lead here was taken by the Stanford heuristic programming group, particularly Buchanan, Feigenbaum, and Shortliffe. What they discovered was that the key to success was to extract from an expert the knowledge he or she used to carry out a specialised task, and then code this knowledge into the computer. In this way they were able to produce 'expert systems' which performed specific tasks at the level of human experts. One of the most important of these early expert systems (MYCIN) was concerned with the diagnosis of blood infections. This system will now be briefly described, and it will then be shown that its implementation led to the problem of how to handle uncertainty in AI.

MYCIN was developed in the 1970s by Edward Shortliffe and his colleagues in collaboration with the infectious diseases group at the Stanford medical school. The medical knowledge in the area was codified into rules of the form: IF such and such is observed, THEN likely conclusion is such and such. MYCIN's knowledge base comprised over 400 such rules which were obtained from medical experts. An example of such a rule will be given in a moment, but first it would be as well to present some evidence of MYCIN's success.

To test MYCIN's effectiveness a comparison was made in 1979 of its performance with that of nine human doctors. The program's final conclusions on ten real cases were compared with those of the human doctors, including the actual therapy administered. Eight other experts were then asked to rate the ten therapy recommendations and award a mark, without knowing which, if any, came from a computer. They were requested to give 1 for a therapy which they regarded as acceptable and 0 for an unacceptable therapy. Since there were eight experts and ten cases, the maximum possible mark was 80. The results were as follows [Jackson, 1986, p. 106]:

MYCIN	52	Actual therapy	46
Faculty-1	50	Faculty-4	44
Faculty-2	48	Resident	36
Inf dis fellow	48	Faculty-5	34
Faculty-3	46	Student	24

So MYCIN came first in the exam, though the difference between it and the top human experts was not significant.

Let us now examine one of MYCIN's rules. The following rule is given by Shortliffe and Buchanan [1975, p. 357]:

If: (1) the stain of the organism is gram positive (S_1), and
(2) the morphology of the organism is coccus (S_2), and
(3) the growth conformation of the organism is chains (S_3)
Then: there is suggestive evidence (0.7) that the identity of the organism is streptococcus (H_1)

In symbols this could be written: If S_1 & S_2 & S_3, then there is suggestive evidence p that H_1, where $p = 0.7$. Here S_1, S_2, S_3 are the observations/symptoms, which support hypothesis H_1 to a particular degree. These rules were obtained from the medical experts. The numbers they contain such as 0.7 were also obtained from the experts. The expert was in fact asked: "On a scale of 1 to 10, how much certainty do you affix to this conclusion?" The answer was then divided by 10.

At first sight it looks as if the figure 0.7 in the rule from MYCIN is an ordinary probability, but this is not the case, as Shortliffe and Buchanan make clear in the following passage [1975, p. 358]:

> ...this rule at first seems to say $P(H_1|S_1\&S_2\&S_3) = 0.7,\ldots$. Questioning of the expert gradually reveals, however, that despite the apparent similarity to a statement regarding a conditional probability, the number 0.7 differs significantly from a probability. The expert may well agree that $P(H_1|S_1\&S_2\&S_3) = 0.7$, but he becomes uneasy when he attempts to follow the logical conclusion that therefore $P(\text{not}.H_1|S_1\&S_2\&S_3) = 0.3$. The three observations are evidence (to degree 0.7) *in favor* of the conclusion that the organism is a streptococcus and should not be construed as evidence (to degree 0.3) *against* streptococcus.

Shortliffe and Buchanan used this observation to motivate the introduction of a *non-probabilistic* model of evidential strength. Their measure of evidential strength was called a *certainty factor*, and certainty factors neither obeyed the standard axioms of probability theory, the Kolmogorov axioms, nor combined like probabilities.

Certainty factors were criticized by those who favoured a probabilistic approach, cf. Adams [1976] and Heckerman [1986], and in fact the next expert system we will consider (PROSPECTOR) did move more in the direction of standard probability.

PROSPECTOR, an expert system for mineral exploration, was developed in the second half of the 1970s at the Stanford Research Institute. A good general account of the system is given by Gaschnig in his 1982. PROSPECTOR's most important innovation was to represent knowledge by an *inference network* (or *net*). This is motivated by Duda *et al.* in their [1976, p. 1076] as follows:

> A collection of rules about some specific subject area invariably uses the same pieces of evidence to imply several different hypotheses. It also frequently happens that several alternative pieces of evidence imply the same hypothesis. Furthermore, there are often chains of evidences and hypotheses. For these

reasons it is natural to represent a collection of rules as a graph structure or *inference net*.

A part of PROSPECTOR's inference network is shown in Figure 1.

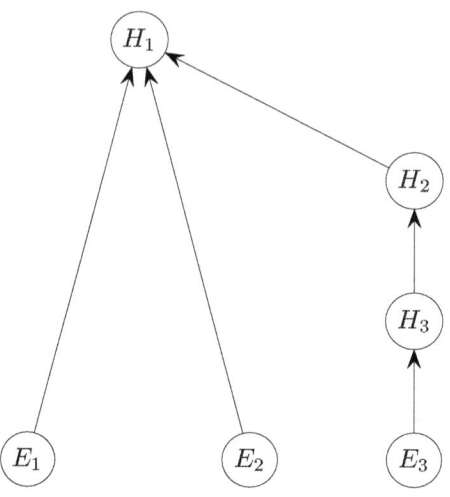

Figure 1.

H_1 = There are massive sulfide deposits.
H_2 = There are clay minerals.
H_3 = There is a reduction process.
E_1 = Barite is overlying sulfide.
E_2 = Galena, sphalerite, or chalcopyrite fill cracks in rhyolite or dacite.
E_3 = There are bleached rocks.

Evidence E_1 is taken as supporting hypothesis H_1, and this is indicated by the arrow joining them in the inference network. Similarly E_2 supports hypothesis H_1, while E_3 supports H_3 which supports H_2 which supports H_1. Note how these rather complicated relations are simply and elegantly represented by the arrows of the network. Each inference arrow has a strength associated with it, and this obtained from the expert as in the case of MYCIN.

PROSPECTOR, however, differs from MYCIN in using subjective Bayesianism rather than certainty factors. This subjective Bayesianism is not entirely pure, since it is combined with fuzzy logic formulae, which were also used in MYCIN. This use of fuzzy logic tended to disappear in further developments.

In PROSPECTOR, Bayesianism is formulated using odds rather than probabilities. The odds on a hypothesis $H[O(H)]$ are defined as follows:

$$O(H) = P(H)/P(\neg H)$$

Writing down Bayes theorem first for H and then for $\neg H$, we get

$$P(H|E) = P(E|H)P(H)/P(E)$$

$$P(\neg H|E) = P(E|\neg H)P(\neg H)/P(E)$$

So dividing gives

(1) $\quad O(H|E) = \lambda(E)O(H)$

where $\lambda(E)$ is the likelihood ratio $P(E|H)/P(E|\neg H)$. (1) is the odds and likelihood form of Bayes theorem, and it is used in PROSPECTOR to change the prior odds on H to the posterior odds given evidence E.

Let us now consider the problems which arise if we have several different pieces of evidence $E_1, E_2, \ldots, E_n$ say. We might in practice have to update using any subset of these pieces of evidence $E_i, E_j, \ldots, E_k$ say, where $(i, j, \ldots k)$ is any subset of $(1, 2, \ldots, n)$. If we use (1), this would involve having values of $\lambda(E_i \& E_j \& \ldots \& E_k)$ for all subsets of $(1, 2, \ldots n)$. When we remember that, on this approach the values of λ are obtained from the domain experts, we can see that obtaining the requisite values of λ is scarcely possible. Clearly some simplifying assumptions are necessary to produce a workable system, and the designers of PROSPECTOR therefore made the following two conditional independence assumptions:

(2) $\quad P(E_1, \ldots, E_n|H) = P(E_1|H) \ldots P(E_n|H)$

(3) $\quad P(E_1, \ldots, E_n|\neg H) = P(E_1|\neg H) \ldots P(E_n|\neg H)$

Given these assumptions, the whole problem of updating with many pieces of evidence becomes simple, and, in fact,

$$O(H|E_1 \& \ldots \& E_n) = \lambda_1 \; \lambda_2 \ldots \lambda_n O(H) \text{ where } \lambda_i = \lambda(E_i)$$

The only remaining problem was whether the conditional independence assumptions (2) and (3) are plausible. The search for a justification of these assumptions led, as we shall see, to the modification of the concept of inference network, and the emergence of the concept of *Bayesian network*.

The concept of Bayesian network was introduced and developed by Pearl in a series of papers: Pearl [1982; 1985a; 1985b; 1986], Kim and Pearl [1983], and a book: Pearl [1988]. An important extension of the theory was carried out by Lauritzen and Spiegelhalter [1988], while Neapolitan's 1990 book gave a clear account of these new ideas and helped to promote the use of Bayesian networks in the *AI* community.

The actual term *Bayesian (or Bayes) network* was introduced in Pearl's [1985b] where it is defined as follows (p. 330):

> Bayes Networks are directed acyclic graphs in which the nodes represent propositions (or variables), the arcs signify the existence of direct causal influences between the linked propositions, and the strengths of these influences are quantified by conditional probabilities.

This verbal account is illustrated by a diagram which is reproduced, with different lettering, in Figure 2.

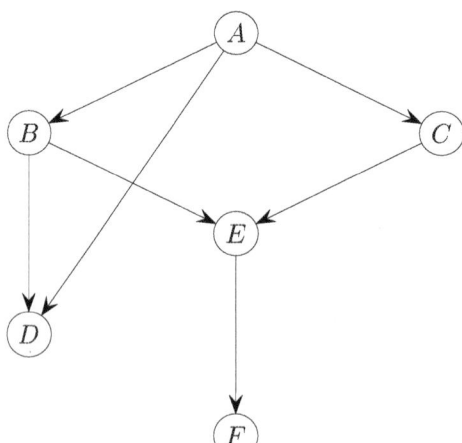

Figure 2.

If we compare the network of Figure 2 with that of Figure 1, two differences should be noted immediately. First of all the arrows in the inference network of Figure 1 represent a relation of support holding between e.g. E_3 and H_3, while the arrows in the Bayesian network of Figure 2 represent causal influences, so that, e.g. the arrow joining A to B means that A causes B. Secondly, corresponding to the first difference, we can say that, in a certain sense, the arrows of a Bayesian network run in the opposite direction to those of an inference network. Pearl puts this point as follows [1986, pp. 253–4]:

> ...in many expert systems (e.g. MYCIN), ...rules point from evidence to hypothesis (e.g. if symptom, then disease), thus denoting a flow of mental inference. By contrast, the arrows in Bayes' networks point from causes to effects or from conditions to consequence, thus denoting a flow of constraints in the physical world.

This reversal of arrows from inference networks to Bayesian networks is illustrated in Figure 3, which shows one pair of nodes taken from the portion of PROSPECTOR's inference network shown in Figure 1.

(a) *Inference Network*

(b) *Bayesian Network*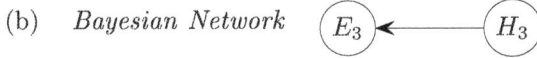

Figure 3. Reversal of Arrows

Here E_3 = There are bleached rocks, while H_3 = There is a reduction process. From the point of view of an inference network (a), we regard the evidence of bleached rocks as supporting the hypothesis that there is a reduction process, while, from the point of view of a Bayesian network (b), we regard there being a reduction process as a cause of there being bleached rocks. In his 1993, Pearl gives an account of his discovery of Bayesian networks, and says that one factor that led him to the idea was his consideration of the concept of influence diagrams introduced by Howard and Matheson (1984). Pearl decided to limit the influences specifically to causal influences. Now Howard and Matheson were working on decision theory. So this is one point where the investigations of decision theory may have had an input into the investigations in artificial intelligence.

I will now make a few further points about Bayesian networks. If, in such a network, an arrow runs from node A to node B, then A is said to be a *parent* of B, and B a *child* of A. Children of A, children of children of A, and so on are known as *descendants* of A. If a node has no parents, it is called a *root*, so that in Figure 2, A is a root. In a Bayesian network, it is possible for a child to have several parents. Thus in Figure 2, E has parents B and C. If, however, every child has at most one parent, the network is called a tree. As in the earlier case of PROSPECTOR's inference networks, in order to make computation feasible, some conditional independence assumptions have to be made. For a Bayesian network, these are that a node is conditionally independent given its parents of the rest of the network except its descendants. I will call the conditional independence assumptions defining a Bayesian network *the generalised Markov condition*.

The nodes of a Bayesian network are random variables. Suppose we specify for each node the conditional probability distribution of that node given its parents, then it follows from the generalised Markov condition that these conditional probability distributions suffice to determine the joint distribution of all the variables of the network. This is an important result since it shows that Bayesian networks enable us to store joint distributions in a very concise way.

After introducing the concept of Bayesian network, Pearl developed algorithms which allow Bayesian updating to take place in such networks. If one of the variables which represents an observation is set to a particular value, the changes brought about by this new information in all the probabilities throughout the tree can be computed in an efficient manner. Pearl began in his 1982 by developing an updating algorithm for a simple form of network, namely a tree. He then extended his algorithm to more complicated networks. Kim and Pearl [1983] generalised from trees to Bayesian networks which are singly connected, i.e. there exists only one (undirected) path between any pair of nodes. Pearl in his 1986 tackled the further extension to Bayesian networks which are multiply connected. This problem was also investigated by Lauritzen and Spiegelhalter who in their 1988 solved it using the idea of reducing a multiply connected network to a tree of cliques. Their algorithm has been generally adopted by the *AI* community.

Let us now turn from these powerful mathematical developments to the consideration of a conceptual point. How exactly are causes and probabilities connected in Bayesian networks? In his original definition which he gave above, Pearl mentions both causes and probabilities. The arrows signify causal influences, while the nodes have associated with them probability distributions conditional on their parents. Pearl's idea about the link between causes and probabilities seems to have been that, if in a network the parents of every node represented the direct causes of that node, then the relevant conditional independence assumptions (the generalised Markov condition) would automatically be satisfied. As he says [1993, p. 52]:

> Causal utterances such as "X is a direct cause of Y" were given a probabilistic interpretation as distinctive patterns of conditional independence relationships that can be verified empirically.

A suggested link between causality and conditional independence in fact goes back to Reichenbach [1956]. Reichenbach considers two events B and C say which are correlated. For example, in a travelling troupe of actors, B = the leading lady has a stomach upset, and C = the leading man has a stomach upset. We can explain such correlations, according to Reichenbach, by finding a common cause, namely that the leading lady and the leading man always have dinner together. The common stomach upsets occur when the food in the local restaurant has gone off. Denote 'dining together' by A. We then have the causal graph shown in Figure 4.

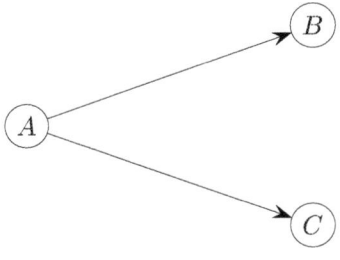

Figure 4.

Reichenbach then claimed that, conditional on A, B and C were no longer correlated but independent, i.e. $P(B\&C|A) = P(B|A)P(C|A)$. He also expressed this idea by saying that a common cause A screens one of its effects B off from the other C. Reichenbach's causal fork is just a simple case of a Bayesian network. We can indeed apply his term 'screening off' to Bayesian networks by saying that in such networks, the parents of a node screen it off from all the other nodes in the network except its descendants.

We are now in a position to summarise the ingenious way in which Bayesian networks solved the problem of handling uncertainty in expert systems. In most of the domains considered, e.g medical diagnosis, a domain expert is very familiar with the various causal factors operating. It should therefore be an easy matter to get him or her to provide a causal network. By the addition of probabilities this can be turned into a Bayesian network. In earlier systems such as MYCIN or PROSPECTOR, conditional independence assumptions were made for the purely *ad hoc* and pragmatic reason of allowing the updating to become possible. For Bayesian networks, however, the causal information obtained from the expert provides a justification for making a set of conditional independence assumptions (the generalised Markov condition) in the manner first suggested by Reichenbach. Moreover as Pearl, Lauritzen and Spiegelhalter have shown, the generalised Markov condition is sufficient to allow Bayesian updating to become computationally feasible. Everything fits together in a most satisfying manner. There is only one weak link in the chain. It turns out that it is possible to have a *bona fide* causal network in which the generalised Markov condition is not satisfied. I have discussed this last point with examples in Gillies [2002], but I will not pursue the development of the theory of Bayesian networks further here. I have given enough of the history of their discovery to enable us to examine in the next three sections the heuristic principles involved.

3 Heuristics Involved: (a) the Use of Philosophical Ideas

The first of the heuristics which I think was involved in the discovery of Bayesian networks was the use of philosophical ideas as a guide to the development of new mathematical concepts. The process which led to the discovery of Bayesian networks was begun by Shortliffe and Buchanan's attempt to construct a formal model for evidential support which could be implemented in their expert system: MYCIN. Shortliffe and Buchanan's key 1975 paper: 'A model of inexact reasoning in medicine' contains 33 references and no less than 14 of these (or over 42%) are to works in the philosophy of science concerned with the confirmation of scientific hypotheses by evidence and related questions concerned with induction and the interpretation of probability. These 14 references are: Barker [1957], Carnap [1950], De Finetti [1972], Harré [1970], Helmer and Rescher [1960], Hempel [1965], Keynes [1921], Popper [1959], Ramsey [1931], Salmon [1966; 1973], Savage [1954], and Swinburne [1970; 1973]. In fact Buchanan and Shortliffe referred to nearly all the philosophers of science who were famous for their works on probability, induction and confirmation.

The main debate within philosophy of science about the confirmation of scientific hypotheses was at the time between the Bayesians and the anti-Bayesians. The Bayesians were divided in turn between the logical Bayesians such as Carnap and the subjective Bayesians such as De Finetti, Ramsey, and Savage. The leading anti-Bayesian was Popper. As we have seen, Shortliffe and Buchanan in constructing their formal model adopted an anti-Bayesian position. They were then immediately attacked by the Bayesians, and it was the members of the subjective Bayesian school, particularly Pearl, who succeeded in developing the successful theory of Bayesian networks. Recently Pearl has introduced some qualifications into his support for Bayesianism. His 2001 paper is significantly entitled: 'Bayesianism and Causality, or Why I am only a Half-Bayesian', but at the very beginning of the paper he reveals that he had no such doubts about the correctness of subjective Bayesianism when he introduced the concept of Bayesian network. This is what he says (2001, p. 19):

> I turned Bayesian in 1971, as soon as I began reading Savage's monograph *The Foundations of Statistical Inference* [Savage, 1962]. The arguments were unassailable: (i) It is plain silly to ignore what we know, (ii) It is natural and useful to cast what we know in the language of probabilities, and (iii) If our subjective probabilities are erroneous, their impact will get washed out in due time, as the number of observations increases.

In other words, Pearl adopted a particular philosophical position (subjective Bayesianism) and this acted as a heuristic guide to his mathematical work.

Pearl may also have been influenced by Reichenbach's philosophical views on causality, for, as we saw earlier, Reichenbach's notion of a causal fork anticipates the concept of Bayesian network in a simple case. However, the textual evidence here is not decisive. In his 1988, Pearl refers to Reichenbach's 1949 book on probability, but this book did not contain a discussion of causal forks which are introduced by Reichenbach in his 1956. In his 1988, Pearl refers to another philosophical work on causality, namely Suppes 1970 monograph: *A probabilistic theory of causation*. However, this work of Suppes does not mention Reichenbach's notion of causal fork.

I think this establishes beyond doubt that philosophical ideas were used as a heuristic guide in the discovery of the mathematical theory of Bayesian networks. But is this an unusual and exceptional case, or does philosophy quite often act as a heuristic in mathematical discovery? The idea that philosophy could be a heuristic guide in the natural sciences is in fact now quite familiar. It was introduced by Popper in 1934 as part of his critique of the Vienna Circle. While the Vienna Circle held that metaphysics was meaningless, Popper argued that metaphysics was not only often meaningful but could be helpful to science. Popper cited the example of atomism which began as a metaphysical theory and long remained one, but which was eventually turned into a scientific theory. Before Popper, Duhem had given many interesting examples of metaphysical ideas acting as heuristics for the development of science. More details of the work of Duhem and Popper on metaphysics in relation to the development of the natural sciences is to be found in Gillies [1993, Chapter 9, Sections 1–3, pp. 189–201].

Although the idea of philosophy acting as heuristic guide is familiar in the case of the natural sciences, there has been surprisingly little discussion of philosophy as a heuristic guide for mathematics. If we examine the history of mathematics, however, we can find many examples of philosophical ideas acting as heuristic guides to mathematical discoveries, though, at the same time, there are also many mathematical discoveries in which philosophy played no role. An obvious example of the influence of philosophy on mathematics is provided by the development of mathematical logic. Frege's revolution in the subject arose from his attempt to support the philosophical view that arithmetic was reducible to logic (see Gillies [1992] for details). The mathematical theory of probability too was strongly influenced by philosophy at earlier periods. The mathematical work of Thomas Bayes was designed to promote Bayesianism which, in turn, was devised in order to answer Hume's sceptical doubts about induction, as I have argued in Gillies [1987]. Another example from probability theory is provided by von Mises who in his development of his frequency theory of probability gave a philosophical analysis and definition of randomness. This definition appeared to have some flaws, and attempts to resolve this difficulty led to important mathematical results by Wald and Church. Details are to be found in Gillies [2000, pp. 105–9].

These examples of the influence of philosophy on the development of mathematics taken from the history of mathematical logic and mathematical probability are not dissimilar from Popper's leading example of the influence of metaphysical ideas on the development of the natural sciences, namely: atomism. Before a precise experimentally testable theory of atomism could be developed it was necessary that atomism as a general view of the world should be elaborated in a less precise, metaphysical fashion. Now logic and probability form an integral part of philosophy because of their importance for epistemology. Some preliminary philosophical analysis of logic and probability was surely necessary to provide a jumping off point for a more precise mathematical theory of these concepts. This explains why in these cases, philosophical ideas were able to act as a guide to mathematical development.

Logic and probability, so I have argued, are part of the subject matter of both philosophy and mathematics. The same is true of the concept of infinity. This is the subject of philosophical disquisitions as well as of Cantor's theory of the transfinite. Indeed Cantor in developing his theory of the transfinite, made an intensive study of philosophical and also theological ideas about the infinite. Details of this are to be found in Dauben's 1979 life of Cantor, which is significantly entitled: Georg Cantor. His Mathematics and Philosophy of the Infinite.

As my final example of philosophical ideas as a heuristic for mathematical discovery, I want to consider a case which is rather different from those of logic, probability, and the infinite. This is Riemann's discovery of non-Euclidean geometry. I have argued that logic, probability and infinity are all subjects of both philosophy and mathematics and that a preliminary qualitative philosophical analysis of these notions was needed before more precise mathematical theories could be developed. Geometry, however, is not *per se* part of philosophy. However, since the time of Plato, geometry has been of great significance for Western philosophy as a prime example of excellent, indeed certain, knowledge, and therefore as a most important example for epistemology. In a famous passage from his 5th Meditation, Descartes says [1641, p. 181]:

> ... I clearly see that existence can no more be separated from the essence of God than can its having its three angles equal to

two right angles be separated from the essence of a [rectilinear] triangle, ...

Now the proposition that the three angles of a rectilinear triangle are equal to two right angles is equivalent to Euclid's 5th postulate. So Descartes is claiming that the truth of Euclidean geometry is as certain as the existence of God. Of course by this he means that the truth of Euclidean geometry is completely certain. Later on Kant claimed that Euclidean geometry was synthetic *a priori*, implying that its truth was known with certainty independently of experience.

These well-known philosophical doctrines affirming the certain truth of Euclidean geometry certainly constituted an obstacle to the discovery of non-Euclidean geometry. Riemann presented his new ideas on non-Euclidean geometry in his famous lecture: 'Über die Hypothesen, welche der Geometrie zu Grunde liegen' (On the Hypotheses which lie at the Foundations of Geometry) delivered as a qualifying lecture (*Habilitationsvorlesung*) for the title of *Privatdozent* to the faculty at Göttingen on 10 June 1854. Riemann regarded it as necessary to begin his lecture with some philosophical analysis. This in effect constitutes an empiricist account of geometry which criticizes implicitly the Kantian view of Euclidean geometry as synthetic *a priori*. Riemann says that he has made use of some philosophical investigations of Herbart, an empiricist philosopher, and he remarks rather modestly [1854, p. 412]:

> ... I think myself the more entitled to ask considerate judgment inasmuch as I have had little practise in such matters of a philosophical nature, where the difficulty lies more in the concepts than in the construction ...

In fact Riemann had studied theology before turning to mathematics and was by no means unfamiliar with philosophy. Here is a passage from his preliminary philosophical discussion [1854, p. 412]:

> ... the propositions of geometry are not derivable from general concepts of quantity ... those properties by which space is distinguished from other conceivable triply extended magnitudes can be gathered only from experience. There arises from this the problem of searching out the simplest facts by which the metric relations of space can be determined, a problem which in nature

of things is not quite definite; for several systems of simple facts can be stated which would suffice for determining the metric relations of space; the most important for present purposes is that laid down for foundations by Euclid. These facts are, like all facts, not necessary but of a merely empirical certainty; they are hypotheses; one may therefore inquire into their probability, which is truly very great within the bounds of observation, and thereafter decide concerning the admissibility of protracting them outside the limits of observation, not only toward the immeasurably large, but also toward the immeasurably small.

The title of Riemann's lecture is itself an implicit criticism of Kant, since Riemann's point is that *hypotheses* (which may be empirically confirmed or disconfirmed) and *not a priori truths* lie at the foundation of geometry. This point is made more explicit in the passage just quoted, since Riemann claims that Euclidean assumptions are 'not necessary but of a merely empirical certainty', and that since 'they are hypotheses', 'one may therefore inquire into their probability'. Riemann regards this probability as very high for what falls within the bounds of observation, but still regards it as possible that Euclidean assumptions might break down 'toward the immeasurably large' or 'toward the immeasurably small'.

More details about Riemann's discovery of non-Euclidean geometry and his empiricism in the philosophy of geometry are to be found in Gillies, [1999, pp. 174–78]. For the purpose of the present paper, however, we can observe that Riemann's empiricist philosophy of geometry, which he developed with the help of Herbart's writings, played a very important role in his discovery of non-Euclidean geometry. It formed the basis of his criticism of the doctrine of Kant and other philosophers who held that Euclidean geometry was known with certainty *a priori*, and so opened up the way to introduce new forms of geometry which contradicted the Euclidean axioms.

In the present section I have given quite a number of examples of mathematical discoveries where philosophical ideas played an important heuristic role. However it should be stressed in conclusion that this is not a general law of mathematical development and there have been many mathematical discoveries in which philosophy played little or not part. An obvious example of such a discovery is the discovery of the concept of group in algebra. This arose from mathematical research into the problem of finding solutions to polynomial equations in terms of radicals. Lagrange found a connection between this problem and permutations of the roots of the equation, and collections of such permutations constituted the first examples of the later concept of abstract group. Here we have a discovery emerging from internal mathematical investigations which did not have a connection with external philosophical questions.

4 Heuristics Involved: (b) New Practical Problems

The study of new practical problems often leads to mathematical discoveries. The discovery of Bayesian networks is a perfect example of this. As we have seen the discovery arose out of the problem of implementing expert systems for medicine, geological exploration and other areas. These expert systems involved handling uncertainty in a way which was rather different from previous applications of the probability calculus. The solution of this problem involved the development of new techniques involving a new mathematical concept.

Once again the pattern here exhibited in the discovery of Bayesian networks is to be found in many other discoveries in the history of mathematics. The mathematical theory of probability itself originated from the problem of calculating fair odds in gambling games. This was a very practical problem at the time, since gambling houses of that period offered odds which were empirically based. A mathematician who could calculate the correct odds stood a good chance of making money. New practical problems about the kinematics and mechanics of moving bodies such as cannonballs, planets or comets stimulated the development of calculus in the 17th century. In the previous section we saw how a philosophical research programme (the attempt to establish logicism in the philosophy of mathematics) led to the development of mathematical logic. However mathematical logic, though it originated in philosophy, was to find practical applications in the field of computer science. The new practical applications led to developments in mathematical logic itself, and, in particular, to the discovery of a quite new type of logic — non-monotonic logic. Some details about the discovery of non-monotonic logic are to be found in Gillies, [1996, pp. 72–75].

Although the investigation of new practical applications often leads to the discovery of new mathematical concepts, sometimes this is not the case because the existing body of mathematics is sufficient for handling the new application. An example of this is provided by Schrödinger's work in quantum mechanics. Schrödinger's equation was a very important discovery in physics, but the equation turned out to be of a type which was familiar to mathematicians, and which could be solved by existing techniques. So, although Schrödinger was investigating some very new, indeed one might almost say, weirdly new phenomena, he was not led to formulating any new mathematical concepts.

Let me conclude this section by comparing the heuristic of using philosophical ideas with that of studying new practical problems. At first sight they seem to be quite distinct and rather opposed approaches. Philosophy, one might think, may be suitable for the abstract pure mathematician like Cantor who is far removed from any practical problem in the real world. Such a person would, it might be thought, be very different from the down to earth researcher working on practical problems. Of course this point of view is correct in Cantor's case, but one finds in many other cases, including our principal example of the discovery of Bayesian networks, that the study of practical problems and philosophical considerations, far from being opposed, actually go hand in hand. The reason for this is that philosophy need not be remote from the real world, but can be closely related to practical action, and, conversely, it may often be difficult to act in practice without some philosophical orientation.

5 Heuristics Involved: (c) Domain Interaction

The third heuristic which I will consider is what I will call: *domain interaction*. This occurs when two separate domains are brought together and partially unified. This process can often result in new discoveries and the growth of knowledge. Domain interaction has been studied by Emily Grosholz, who has emphasized its role in the development of mathematics. In this section, therefore, I will reverse the order used in the two preceding sections. I will first give a general account of Grosholz's ideas on domain interaction, including examples of where it has led to mathematical discovery. I will then show that domain interaction was an important heuristic principle involved in the discovery of Bayesian networks. In fact the example of Bayesian networks provides a striking vindication of Grosholz's ideas on this subject.

In a series of publications [1981; 1985; 1991; 1992], Grosholz has studied a number of cases in which knowledge (particularly mathematical knowledge) has advanced through the interaction of separate domains. In 1981, she considers Logic and Arithmetic, in 1985 Logic and Topology, while in her 1992 she argues that Leibniz invented and developed the calculus by bringing together geometry, algebra, number theory, and mechanics. Her 1991 book shows that to a remarkable extent all Descartes' intellectual work can be seen as bringing together different domains. As she says [1991, pp. 2-3]:

> ... Cartesian domains ... can be understood as a novel amalgamation of formerly distinct or at least very incompletely unified domains: the *Geometry* brings together geometry and algebra, the *Principles* geometry and physics, the *Treatise of Man* physics and medical physiology, and the *Meditations* mechanical philosophy and scholastic theology.

This is an interesting passage since it shows that the heuristic of domain interaction is not limited to mathematics, but applies to other subjects as well. However the passage also gives one of the most famous examples of domain interaction in mathematics, namely the bringing together of geometry and algebra to create analytic geometry. Although Grosholz approves of Descartes's method of bringing together separate domains, she nonetheless criticizes the way in which he carries out this process. In her view the interaction of different domains is most fruitful, if, while interacting, they nonetheless retain some degree of autonomy. An attempt to reduce one domain to the other will generally inhibit fruitful developments. As she says [1991, p. 3]:

> ... the unification of domains contributes to the growth of knowledge when and because it exploits partially shared structure between domains that none the less retain their autonomy and distinctness. Revelation is impaired when domains are held too far apart, or assimilated too closely. But Descartes's way of constructing knowledge can produce both these unfortunate outcomes ...

According to Grosholz, Leibniz was more successful that Descartes in handling domain interaction (see [Grosholz, 1992]).

Another important concept which Grosholz uses in this connection is the concept of hybrid. As she says [2000, p. 82];

Moreover, my examination of the growth of mathematical knowledge sheds important light on mathematical hybrids, objects which exist in the overlap of domains and provoke discovery in unexpected ways.

An important feature of such hybrids is that they exhibit a kind of instability or inconsistency. As Grosholz says [2000, p. 88]:

> ...the two domains as it were overlap, or are superimposed. At this overlap, objects are constituted which must simultaneously exhibit features of both domains; if the domains are truly heterogeneous, one must expect a kind of submerged heterogeneity in these objects. And in fact such hybrids often exhibit an instability or inconsistency that is however held in place or made tractable by the rational relatedness provided by the abstract structure that holds the domains together.

This instability or inconsistency is not seen by Grosholz as a defect, but rather as a potential stimulus to further growth and development.

Let me now show that these ideas apply very well to the example of Bayesian networks. In fact, Bayesian networks involved two instances of domain interaction. To begin with, Bayesian networks put together the domains of probability theory and graph theory which had previously been largely separate. Secondly, however, Bayesian networks put together the domains of probability theory and causality. In fact there had earlier been the beginning of an attempt in the philosophy of science community to connect these domains. Suppes [1970] *A probabilistic theory of causation* is a leading example of this trend. However the development of the concept of Bayesian network was a notable advance in linking the two domains. In a Bayesian network, an arrow joining two nodes A and B usually indicates that there is a causal connection between A and B. Furthermore each node in a Bayesian network has a conditional probability distribution associated with it. Thus causality and probability are brought together.

However this hybrid of causality and probability is by no means unproblematic. Pearl originally hoped that the causal connections between the nodes of a network would justify adopting the generalised Markov condition for the probability distributions. However it emerged that there can be genuine causal graphs for which the generalised Markov condition does not hold. More details about this are to be found in Gillies [2002]. So the relations between causality and probability in a Bayesian network turn out to be highly problematic. Bayesian networks thus fit very well the descriptions which Grosholz gives of other mathematical hybrids. As she says [2000, p. 88]: ' ...one must expect a kind of submerged heterogeneity in these objects. And in fact such hybrids often exhibit an instability or inconsistency ...'

6 Heuristics of Mathematical Discovery versus Logic of Mathematical Discovery

Having given my example of a recent mathematical discovery and attempted to analyse the heuristics which were involved, I now want to raise the general question of whether such heuristics constitute a kind of generalised logic so that one could speak of a logic of mathematical discovery, or whether heuristic principles are not logical in character. This question is by no means an easy one. The core of logic is obviously standard deductive logic. However, it has often been suggested that logic could be extended to include not just deductive inferences but ampliative inferences of various kinds. For example, many philosophers of science have supported the idea of an inductive logic. Might heuristic principles constitute an extension of logic of which inductive logic is just a part?

I will begin my examination of this problem by considering an interesting related discussion by Ladislav Kvasz in his 2002. Kvasz here deals not with the relation between heuristics and logic, but with the obviously closely connected question of the relation between dialectics and logic. Kvasz in his paper criticizes dialecticians, but under that heading he includes not just the paradigm dialecticians (Hegel and the Marxists), but also Popper and Lakatos whom he regards as also dialecticians, notwithstanding their striking disagreements with Hegel and the Marxists. What is common to all these thinkers according to Kvasz is that they regard dialectics as a branch of logic. As Kvasz himself says [2002, p. 211]:

> Usually, the dialecticians believe that the pattern of the development of knowledge is of a logical nature (Hegel's idea of dialectical logic, Popper's logic of scientific discovery, or Lakatos' logic of mathematical discovery), which creates a tension between the development of knowledge and formal logic.

This 'confusion of dialectic with logic' [Kvasz, 2002, p. 211] is responsible, according to Kvasz, for grave failings in Hegel on the one hand and in Popper and Lakatos on the other. However, these failings are different in the two cases. The problem with Hegel and the Hegelians is that they regard their dialectical logic as being in competition with and superior to ordinary deductive logic. Hegelians therefore reject ordinary deductive logic which Kvasz thinks is a mistake. Popper and Lakatos did not give up ordinary deductive logic, but their attempt to reconcile it with the development of knowledge led to them confining their analyses to cases in which the conceptual changes in the growth of knowledge are relatively small. This is how Kvasz puts this argument [2002, p. 229]:

> Both solutions to the dialectician's conflict between logic and evolution of knowledge are unsatisfactory. Philosophers who follow Hegel, in the attempt to replace classical logic by some new dialectical one, were unable to offer anything comparable to the successive formal logic, and thus their research programme degenerated. On the other hand, dialecticians like Popper or Lakatos, who were not prepared to sacrifice logic, and thought that logical consistency is crucial to rational discourse, were forced to give up evolution. The fact that Lakatos was unable to reconstruct any deeper conceptual change in history of mathematics or physics is not accidental. As a dialectician, he conceived evolution to be in conflict with logic, but as Popper's disciple he was not prepared to give up logic. Thus he omitted some of the most interesting moments in the history of mathematics. If he had tried to reconstruct them, he would have been forced to violate logic. Therefore he reconstructed only those changes, in which relatively small conceptual changes occur. ...
>
> The one extreme is *dialectical logic* (of Hegel and Marxism), which for the sake of evolution sacrifices logic. The other extreme is *logical dialectic* (of Popper or Lakatos), which for the sake of logic sacrifices evolution.

Kvasz argues for this general position by giving an analysis of some changes in the development of mathematics which he regards as being too large to be compatible with formal logic. These changes all involve a change in the form of the language used. Following Wittgenstein in the *Tractatus*,

Kavasz regards any language (L say) as having a form which is not expressible in the language. We can however incorporate the form of the language L into L thereby creating a new language L' say. A simple example of this process occurred in the transition between the language of perspective used by Renaissance painters and the language of projective geometry created by Desargues. As Kvasz says [2002, p. 221]:

> ...the centre of projection represents, in an abstract form, the eye of the painter from Dürer's drawing. For Desargues, ..., the point of view is explicitly incorporated into language.

In fact Kvasz analyses a whole series of examples of changes in mathematics which follow this pattern in his papers [1998] and [2000].

We can now see clearly why formal logic is inadequate to deal with such changes. Any system of formal logic presupposes a language L in which it is formulated, and this language is held constant when the deductions are being made. If therefore we make a fundamental change in the character of the language, altering it from L to L', this change cannot be captured using formal logic. On the other hand we can apply formal logic without any problems either within L or within L' so that there is no need to abandon formal logic altogether as the Hegelians deem to be necessary. Formal logic has only to be given up temporarily in the course of a large change involving a considerable alteration in the form of the language used. This then is a brief summary of Kvasz's position. Let us now see if we can apply it to our problem about heuristics and logic.

It is clear that the discovery of Bayesian networks involved the creation of a new language formed through the synthesis of the languages of earlier probability theory and graph theory. The language of Bayesian networks with its network diagrams has an iconic character which is not to be found in earlier probability theory. As this is a major change in language, then we can use Kvasz's argument to conclude that the discovery of Bayesian networks is a transition which cannot be explicated logically so that the heuristics involved are not logical in character. Indeed we can generalise to say that many applications of the domain interaction heuristic take us outside logic. Cartesian geometry, for example, has its own specific language which differs both from the language of classical Euclidean geometry and from that of algebra unrelated to geometry. Similarly calculus introduced new symbolisms such as dy/dx or $\ddot{y}$ which made the language of calculus radically different from preceding mathematical languages.

Changes of this magnitude cannot, according to Kvasz's argument, be explicated using logic. However, Kvasz's analysis also indicates that some changes might be logically explicated. These would be smaller changes. An example might be the discovery of the proof of a mathematical conjecture where both the conjecture and the subsequent proof are formulated within a well-defined mathematical system, which is not changed in the process of discovery. There is no reason why the heuristics of discoveries of this sort might not be explicated in a way that could be described as logical.

These conclusions are supported by another approach to the problem. This approach relies on the connection between logic and mechanisation. If mathematical proofs are translated into formal logic then the validity of each step can be checked mechanically by means of a computer. The discovery of the proof, however, can be left entirely in the hands of human mathematicians. The development of automated theorem proving, and of non-monotonic logic programming languages such as PROLOG has carried the mechanisation process one stage further by mechanising the construction of proofs. In this respect, then, it goes beyond Fregean formal logic.

I have suggested (in [Gillies, 1996, p. 85]) a way of characterising this new kind of logic which has been introduced by investigations into artificial intelligence. The formula proposed is

$$Logic = Inference + Control$$

When we employ Logic, we start with a set of assumptions from which we want to derive some conclusions. To carry out these derivations we need a set of rules of inference (the Inference component). If the derivation is carried out by a trained mathematician, then he or she will rely on intuition to decide which rule of inference to use at a particular stage in order to carry out the derivation. If, however, we are trying to program a computer to carry out the derivation, then we will have to give the computer guidance as to which assumptions to choose and which rules of inference to apply. This guidance constitutes the Control component. Thus the Control component might specify at each stage of the derivation, which of the assumptions should be employed, and which of the rules of inference should be applied to these assumptions or to previously obtained results. More generally, the Control component would be designed to help in the construction of a derivation or proof of a conclusion.

I further suggested (in [Gillies, 1996, Ch. 5, pp. 98–112]) that this formula enables one to defend the possibility of an inductive logic. The development of machine learning has lead to the formulation of inductive rules of inference, while confirmation theory constitutes the control component. In the case of automated theorem proving, the heuristics used could be formulated as part of the control component, and could then, using the formula above, be considered part of a logic of mathematical discovery. This criterion suggests therefore that a heuristic can be considered a logical principle if it can be formulated with precision and incorporated into a successful computer system for automated theorem proving. This criterion implies the Kvasz criterion since, at least as things stands at present, any automated theorem proving system has to operate within a fixed formal language specified at the beginning.

The kind of heuristics which I have considered in this paper (use of philosophical ideas, consideration of new practical problems, and domain interaction) are to vague in character to be suitable for precise formulation and implementation in programs for automated theorem proving. I would therefore argue that they are not logical in character.

A critic might say at this point that heuristics which are not precise enough to become part of logic are unlikely to provide much useful guidance. However such a comment would be unfair. The somewhat vague non-logical heuristics considered in this paper are certainly not precise enough to guide a computer in the execution of a program. However they are precise enough to suggest strategies for a human mathematician carrying out mathematical research. Moreover the kind of strategies suggested by the three heuristics given are rather different from those commonly adopted by human mathematical researchers. It is all too common for research mathematicians to become exclusively absorbed in their own small field and to devote themselves to reading only the literature of that specialty. The analysis given of the discovery of Bayesian networks suggests a quite different sort of research strategy, one which would involve a broader more interdisciplinary approach, with the study of some philosophy, an interest in areas which might require new techniques for successful practical applications, and a knowledge of several branches of mathematics which could be brought together for 'domain interaction'. The example even suggests some more specific recommendations. Mathematicians preparing for research in the area of probability and statistics would normally take a master's degree in this speciality. The discovery of Bayesian networks suggests that it might be worth including a course on the philosophical and foundational aspects of probability and statistics as part of the preparation of the future researcher. Yet this is rarely if ever done. In effect the heuristics considered in this paper, though not precise enough to guide a computer, do definitely suggest strategies for humans who want to carry out research in mathematics.

BIBLIOGRAPHY

[Adams, 1976] J.B. Adams. A probability model of medical reasoning and the MYCIN model. *Mathematical Biosciences*, **32**, pp. 177–86, 1976.

[Barker, 1957] S.F. Barker.) *Induction and Hypothesis: A Study in the Logic of Confirmation.* Ithaca, New York, Cornell University Press, 1957.

[Buchanan:Shortliffe, 1984] B.G. Buchanan and E.H. Shortliffe (eds). *Rule-based expert systems: the MYCIN experiments of the Stanford heuristic programming project.* Reading, Mass, Addison-Wesley, 1984.

[Carnap, 1950] R. Carnap. The two concepts of probability. In *Logical Foundations of Probability*, Chicago, University of Chicago Press, pp. 19–51, 1950.

[Cellucci, 2002] C. Cellucci. *Filosofia e matematica.* Roma-Bari, Laterza, 2002.

[Corfield, 2003] D. Corfield. *Towards a Philosophy of Real Mathematics.* Cambridge, Cambridge University Press, 2003.

[Dauben, 1979] J.W. Dauben. *Georg Cantor. His Mathematics and Philosophy of the Infinite.* Cambridge, Massachusetts and London, England, Harvard University Press, 1979.

[Descartes, 1641] R. Descartes. *Meditations on First Philosophy*, 1641. English translation in *The Philosophical Works of Descartes*, Volume 1, translated by Elizabeth S. Haldane and G.R.T. Ross, Cambridge, Cambridge University Press, pp. 131–99, 1970.

[De Finetti, 1972] B. De Finetti. *Probability, Induction, and Statistics — the Art of Guessing*, New York, Wiley, 1972.

[Duda, Hart and Nilsson, 1976] R.O. Duda, P.E. Hart and N.J. Nilsson. Subjective Bayesian methods of rule-based inference systems. In *Proceedings of the National Computer Conference (AFIPS)*, **45**, pp. 1075–82, 1976.

[Gaschnig, 1982] J. Gaschnig. Prospector: an expert system for mineral exploration. In Donald Michie (ed.) *Introductory readings in expert systems*, New York, Gordon and Breach, pp. 47–64, 1982.

[Gillies, 1987] D.A. Gillies. Was Bayes a Bayesian? *Historia Mathematica*, **14**, pp. 325–46, 1987.

[Gillies, 1992] D.A. Gillies. The Fregean Revolution in Logic. In Donald Gillies (ed.) *Revolutions in Mathematics*, Oxford, Oxford University Press, pp. 265–305, 1992.

[Gillies, 1993] D.A. Gillies. *Philosophy of Science in the Twentieth Century.* Oxford UK & Cambridge USA, Blackwell, 1993.

[Gillies, 1996] D.A. Gillies. *Artificial Intelligence and Scientific Method*. Oxford, Oxford University Press, 1996.

[Gillies, 1999] D.A. Gillies. German Philosophy of Mathematics from Gauss to Hilbert. In Anthony O'Hear (ed.), *German Philosophy since Kant*, Cambridge, Cambridge University Press, pp. 167–92, 1999.

[Gillies, 2000] D.A. Gillies. *Philosophical Theories of Probability*, London and New York, Routledge.

[Gillies, 2002] D.A. Gillies. Causality, Propensity, and Bayesian Networks. *Synthese*, **132**, pp. 63–88, 2002.

[Grosholz, 1981] E.R. Grosholz. Wittgenstein and the Correlation of Logic and Arithmetic. *Ratio*, **23**, pp. 31–42, 1981.

[Grosholz, 1985] E.R. Grosholz. Two Episodes in the Unification of Logic and Topology. *British Journal for the Philosophy of Science*, **36**, pp. 147–57, 1985.

[Grosholz, 1991] E.R. Grosholz. *Cartesian Method and the Problem of Reduction*. Oxford, Oxford University Press, 1991.

[Grosholz, 1992] E.R. Grosholz. Was Leibniz a Mathematical Revolutionary? In Donald Gillies (ed.) *Revolutions in Mathematics*, Oxford, Oxford University Press, pp. 117–133, 1992.

[Grosholz, 2000] E.R. Grosholz. The Partial Unification of Domains, Hybrids, and the Growth of Mathematical Knowledge. In Emily Grosholz and Herbert Breger (eds.) *The Growth of Mathematical Knowledge*, Dordrecht, Boston, London, Kluwer, pp. 81–91, 2000.

[Harré, 1970] R. Harré. Probability and confirmation. In *The Principles of Scientific Thinking*, Chicago, University of Chicago Press, 1970.

[Heckerman, 1986] D. Heckerman. Probabilistic interpretations for MYCIN's certainty factors. In L.N. Kanal & J.F. Lemmer (eds.) *Uncertainty in Artificial Intelligence*, Amsterdam, North-Holland, pp. 167–96, 1986.

[Helmer and Rescher, 1960] O. Helmer and N. Rescher. On the epistemology of the inexact sciences. *Project Rand R-353*, February 1960.

[Hempel, 1965] C.G. Hempel. Studies in the logic of confirmation. In *Aspects of Scientific Explanation and other Essays in the Philosophy of Science*, New York, The Free Press, pp. 3–51, 1965.

[Howard and Matheson, 1984] R.A. Howard and J.E. Matheson. Influence diagrams. In R.A.Howard and J.E. Matheson (eds.) *The principles and applications of decision analysis*, Vol. 2, Menlo Park, CA, Strategic Decisions Group, pp. 721–62, 1984.

[Jackson, 1986] P. Jackson. *Introduction to expert systems*. Wokingham, England, Addison-Wesley, 1986.

[Keynes, 1921] J.M. Keynes. *A Treatise on Probability* 1921. New York, Harper and Row, 1962.

[Kim and Pearl, 1983] J.H. Kim and J. Pearl. A computational model for combined causal and diagnostic reasoning in inference systems. *Proceedings of the 8th International Joint Conference on AI (IJCAI-85)*, pp. 190–3, 1983.

[Kvasz, 1998] L. Kvasz. History of Geometry and the Development of the Form of its Language. *Synthese*, **116**, pp. 141–86, 1998.

[Kvasz, 2000] L. Kvasz. Changes of Language in the Development of Mathematics. *Philosophia Mathematica*, **8**, pp. 47–83, 2000.

[Kvasz, 2002] L. Kvasz. Lakatos' Methodology Between Logic and Dialectic. In George Kampis, Ladislav Kvasz and Michael Stöltzner (eds.) *Appraising Lakatos. Mathematics, Methodology and the Man*, Vienna Circle Library, Dordrecht, Boston, London, Kluwer, pp. 211–41, 2002.

[Lauritzen and Spiegelhalter, 1988] S.L. Lauritzen and D.J. Spiegelhalter. Local computations with probabilities on graphical structures and their application to expert systems (with discussion). *Journal of the Royal Statistical Society B*, **50**, pp. 157–224, 1988.

[Neapolitan, 1990] R.E. Neapolitan. *Probabilistic reasoning in expert systems. Theory and algorithms*. New York, John Wiley, 1990.

[Ng and Abrahamson, 1990] K. Ng and B. Abrahamson. Uncertainty management in expert systems. *IEEE Expert*, **5**, pp. 29–48, 1990.

[Pearl, 1982] J. Pearl. Reverend Bayes on inference engines: a distributed hierarchical approach. *Proceedings of the National conference on AI (ASSI-82)*, pp. 133–6, 1982.

[Pearl, 1985a] J. Pearl. How to do with probabilities what people say you can't. *Proceedings of the Second IEEE Conference on AI Applications*, Miami, Fl., pp. 6–12, 1985.

[Pearl, 1985b] J. Pearl. Bayesian networks: a model of self-activated memory for evidential reasoning. *Proceedings of the Cognitive Science Society*, Ablex, pp. 329–34, 1985.

[Pearl, 1986] J. Pearl. Fusion, propagation and structuring in belief networks. *Artificial Intelligence*, **29**, pp. 241–88, 1986.

[Pearl, 1988] J. Pearl. *Probabilistic reasoning in intelligent systems. Networks of plausible inference*. San Mateo, California, Morgan Kaufmann, 1988.

[Pearl, 1993] J. Pearl. Belief networks revisited. *Artificial Intelligence*, **59**, pp. 49–56, 1993.

[Pearl, 2001] J. Pearl. Bayesianism and causality, or, why I am only a half-Bayesian. In David Corfield and Jon Williamson (eds.) *Foundations of Bayesianism*, Dordrecht, Boston, London, Kluwer, pp. 19–36, 2001.

[Popper, 1959] K.R. Popper. Corroboration, the weight of evidence, and statistical tests. In *The Logic of Scientific Discovery*, New York, Scientific Editions, pp. 387–419, 1959.

[Ramsey, 1931] F.P. Ramsey. *The Foundations of Mathematics and other Logical Essays*. London, Kegan Paul, 1931.

[Reichenbach, 1949] H. Reichenbach. *Theory of Probability*. Berkeley, University of California Press, 1949.

[Reichenbach, 1956] H. Reichenbach. *The Direction of Time*. Berkeley, University of California Press, 1956.

[Riemann, 1854] B. Riemann. On the Hypotheses which lie at the Foundations of Geometry, 1854. English translation in David E. Smith (ed.), *A Source Book of Mathematics*, Volume Two, New York, Dover Publications, pp. 411–25, 1959.

[Salmon, 1966] W.C. Salmon. *The Foundations of Scientific Inference*. Pittsburgh, Pennsylvania, University of Pittsburgh Press, 1966.

[Salmon, 1973] W.C. Salmon. Confirmation. *Scientific American*, May 1973, pp. 75–83, 1973.

[Savage, 1954] L.J. Savage. *The Foundations of Statistics*. New York, Wiley, 1954.

[Savage, 1962] L.J. Savage (ed). *The Foundations of Statistical Inference*. London, Methuen, 1962.

[Shortliffe and Buchanan, 1975] E.H. Shortliffe and B.G. Buchanan. A model of inexact reasoning in medicine. *Mathematical Biosciences*, **33**, pp. 351–79, 1975.

[Suppes, 1970] P. Suppes. *A probabilistic theory of causation*. Amsterdam, North Holland, 1970.

[Swinburne, 1970] R.G. Swinburne. Choosing between confirmation theories, *Philosophy of Science*, **37**, pp. 602–13, 1970.

[Swinburne, 1973] R.G. Swinburne. *An Introduction to Confirmation Theory*. London, Methuen, 1973.

Do We Really Need Axioms in Mathematics?

GIANLUIGI OLIVERI

1 Introduction

Many mathematical theories began their life as informal theories, axiomatization came about only later. At times the process of axiomatization happened shortly after the introduction of the theory, as in the case of set theory, but at other times, as in the case of arithmetic, it took a long while before an acceptable set of axioms was provided.

But regardless the exact timing of its introduction, we can certainly say that the axiomatic method has had a profound and lasting influence on mathematics so much so that Hilbert did not hesitate to speak of 'axiomatic thinking' [Hilbert, 1918].

To give an explanation of what Hilbert meant by 'axiomatic thinking', we ought to consider that, according to him, given a mathematical theory T, it is possible to distinguish between results obtained within it which are of central importance — like the Fundamental Theorem of Calculus, the Fundamental Theorem of Algebra, the well-ordering theorem in set theory, the Completeness Theorem for first-order logic — and those which are not so important.

Relatively to these fundamental results, the theory can be developed either *downwards* or *upwards*. Developing the theory downwards means deriving consequences from the fundamental results. Developing the theory upwards means finding some statements of the theory from which it is possible to derive the fundamental results. Axiomatic thinking is, for Hilbert, the way of regarding a mathematical theory from the point of view of what I am going to call 'upward continuation'.

I must here emphasize that the influence exercised by the axiomatic method in mathematics is not confined to representing mathematical theories as conceptual constructions erected on foundations exemplified by a set of axioms. In fact, the influence exercised by the axiomatic method in mathematics extends also to heuristics.

There are many cases in mathematics in which the application of the axiomatic method, and the study of problems connected with it, have led to the discovery of new mathematical results. Trivially, the very act of singling out and formulating a set of axioms for a given mathematical theory represents already the discovery of a new result.

Other examples of genuine discoveries obtained through an application of the axiomatic method, or through the study of problems connected with its applications, are: the independence of the Axiom of Choice (**AC**) from **ZF**[1] and of the Continuum Hypothesis (**CH**) from **ZFC**, the non-Euclidean geometries, the finite geometries, the emergence of structure — number theoretical, algebraic, topological, etc. — as an object of mathematical investigation, and, of course, Gödel's incompleteness theorems together with all the results belonging to what we may call 'theory of formal systems'.

Lastly, the axiomatic method has also affected the development of mathematics. Indeed, if we look at the historical development of mathematical theories, we realize that until the XVIIIth century the only axiomatized mathematical theory was Euclidean geometry; yet, nowadays, within classical mathematics, apart from the possibility of translating/reducing mathematical theories to an axiomatized version of set theory, mathematical theories are born with specific axioms of their own.

But, if it is correct to say that in the past two centuries the axiomatic method has taken mathematics by storm, it is also true that in the first half of the past century a number of important limiting results were established.

However, in spite of these limiting results, the axiomatic method still seems to represent one of the most important paradigms adopted by mainstream classical mathematics. This paper intends to investigate some of the reasons behind such a state of affair with the view of clarifying the problem concerning whether we really need axioms in mathematics.

2 Formal and informal mathematics: axiomatic vs. intuitive/genetic?

The axiomatic method in contemporary mathematics is usually associated with the formulation and development of so-called 'formal theories', that is,

---

[1] This result was expected by Zermelo. See on this [Zermelo, 1908], §2a, p. 187.

with the formulation and development of theories obtained by adding to a system of logic, say, first-order logic, terms, axioms, primitive notions, and definitions of a given mathematical theory.

Formal theories are, of course, contrasted with informal ones. But what do we have to understand by 'informal mathematical theory' or by 'informal mathematics'? Perhaps, the best way of proceeding in the attempt to answer this question is by way of examples.

One of the best known examples of informal mathematics present in the philosophical literature is that given by Plato in the *Meno*, where a slave boy, under Socrates's unrelenting prodding, solves the following problem: given a square S, find a square S_1 the area of which measures double the area of S. Another well known example of informal mathematics is that provided by Lakatos in *Proofs and Refutations*. It is particularly instructive to examine the case of the slave boy.

First of all, Socrates, in his *experiment* aimed at showing that knowledge is recollection, nowhere provides the definitions of 'square', 'diagonal', 'area', etc. but, with the help of diagrams, the relevant properties of which are taken as being self-evident, produces what we might call 'ostensive definitions' of some of the above mentioned notions in which the appeal to intuition is essential as well as obvious:[2]

> *Socrates.* Now boy, you know that a square is a figure like this? (*Socrates begins to draw figures in the sand at his feet. He points to the square* ABCD).
>
> *Socrates.* And these lines which go through the middle of it are also equal?

Furthermore, the slave boy's painful *route to recollection* consists of a process of conjectures and refutations based on the intuitive properties of the diagrams produced, a process of conjectures and refutations which eventually culminates into a proof.

In this story there are several features of interest for us. First, in the informal approach followed by Socrates we have the account of 'the genesis' of the notions which eventually lead to the solution of the problem. And if one were to take this as the paradigm of how mathematical activity takes

---

[2][Plato, 1956], 82 b 9-12 and 82 c 3-4.

place, he should then conclude that mathematical activity consists in generating, inventing or discovering *by means of intuition* the notions which enable us to solve the problems of the theory.

Secondly, the use of intuition is all pervading, given the eminently diagrammatic approach to definition and proof in which the relevant properties of the diagrams are taken to be self-evident.

Thirdly, the fact that the slave boy story includes, besides the solution of the problem, also all the other aborted attempts at a solution, seems to show that the process of growth of mathematical knowledge consists in a dialectic of proofs and refutations, a dialectic which highlights both the so-called contexts of 'discovery' and 'justification'.

If we look, instead, at a possible formal approach to the problem proposed by Socrates, a formal approach which consists in: (1) adding to a system of logic terms, axioms, primitive notions and definitions of Euclidean geometry, (2) proving Pythagoras theorem from them, and (3) using Pythagoras theorem to solve the problem, we realize that this greatly differs from the informal one.

For, in the formal approach there seems to be no appeal to intuition, because definitions, axioms and rules of transformation are clearly laid out from the beginning, and the proof produced appeals only to the meaning of the axioms, definitions and rules of transformation.

Moreover, the often counter-intuitive nature of formal proofs gives little indication concerning the so-called 'context of discovery' and, therefore, it appears to produce only a one sided picture of mathematical activity in which this is seen exclusively as a process of justification.

But, given the considerations above, it is now legitimate to ask what drives mathematicians to produce formal theories.

3 On some of the aims of the axiomatic method

One of the earliest aims which motivated the introduction of the axiomatic method in mathematics was that of providing a *foundationalist* justification for the assertion of mathematical statements. But what does this mean?

Given a mathematical theory, say, Euclidean geometry, the Greeks distinguished between statements of the theory which are self-evidently true (false) and those which are not. Clearly, the statements which need justifications in terms of proof, rather than in terms of an elucidation of their meaning, are those which are not self-evidently true (or false).

Since proving a statement A means showing that A is a logical consequence of a set of assumptions $\{A_1, \ldots, A_n\}$, it is obvious that, unless we manage to find self-evidently true statements $A_1, \ldots, A_n$ from which to prove A, an infinite regress would start in the process of justification which would ultimately allow us to assert only a conditional statement of the kind 'If $A_1, \ldots, A_n$ are all true also A is true'.

For the Greeks, the axioms and postulates of Euclidean geometry were self-evidently true statements and, therefore, provided an adequate foundation for the assertion of statements belonging to Euclidean geometry.

With the advent of the non-Euclidean geometries the idea that an axiom had to be a self-evidently true statement largely faded away. And the assertion of mathematical statements received a *coherentist* foundation, that is, the assertion of a mathematical statement A was considered to be justified if and only if it could be shown that A was a logical consequence of a consistent set of assumptions.

Although this state of affairs undermined the rôle of proof as what shows that mathematical statements are true, it did not compromise the use of the axiomatic method in mathematics. For, if the Euclidean postulate of parallels is not self-evidently true, and is actually independent of the other axioms of Euclidean geometry, the way is open to the proliferation of axiom systems.

Another reason behind the irresistible push towards axiomatization in mathematics has been the increasing scepticism about intuition as a means for attaining truth. Such an attitude towards intuition gained momentum in the XIXth and in the XXth centuries as a consequence, on the one hand, of the large number of important cases in which intuition had led mathematicians astray and, on the other hand, of the success achieved by formal theories in sorting out, or barring out, the troubles which had beset their informal counterparts. One of the best known cases of this kind is that of the introduction of infinitesimals in mathematical analysis at the hands of Newton and Leibniz. Other famous cases are those of Euler's formula for polyhedra $\mathbf{V} - \mathbf{E} + \mathbf{F} = 2$, of Euclid axiom 5 according to which the whole is greater than the part, and of the unrestricted principle of comprehension in naïve set theory (**NST**).

A third reason which has strongly favoured the process of axiomatization of mathematical theories has been the attempt to justify the proof of results central to a particular mathematical theory showing that the new and, of course, counterintuitive principles used in the proof are not gratuitous, *ad hoc* assumptions, but are part of a consistent set of principles from which we can obtain the known results of the theory as well as the central results in question. A classical example of this kind is given by the story of the acceptance of Zermelo's proof of the well-ordering theorem.

As is well known, in 1904 Zermelo produced a 'Proof that every set can be well-ordered'.[3] His proof used a then new principle, the Axiom of Choice, which immediately became the object of a great controversy among mathematicians.

In 1908 Zermelo published 'A new proof of the possibility of a well-ordering'[4] and 'Investigations in the foundations of set theory I'.[5] In these papers Zermelo (i) discussed the objections which had been directed against his first proof [Zermelo, 1908]; (ii) produced a new proof of the well-ordering theorem [Zermelo, 1908]; and (iii) provided an axiomatization of set theory which represented both an ideal background for his proof of the well-ordering theorem and an effective barrier against the known set theoretical paradoxes [Zermelo, 1908a].

Now, certainly, Zermelo's detailed answers to the various objections to his first proof in [Zermelo, 1908] played a rôle in the acceptance of his second proof of the well-ordering theorem. But it seems to me that the most convincing argument provided by Zermelo in his 1908 papers consisted in showing that his proof of the well-ordering theorem was part of a new system of set theory, **Z**, which differed from and was better than **NST**.

4 Against the axiomatic method

In spite of the great success enjoyed by the axiomatic method in contemporary mathematics, there have been those who have developed a rather sceptical attitude towards it. In some cases, such an attitude manifested itself in a considerable number of serious objections some of which I will examine in what follows.

Against the idea that mathematical theories must be axiomatized, Brouwer observes that axiomatic systems set arbitrary and unacceptable limitations to the creative activity of the subject, creative activity which, according to Brouwer, 'develops in self-unfolding guided by free arbitrariness'[6].

---
[3] [Zermelo, 1904].
[4] [Zermelo, 1908].
[5] [Zermelo, 1908a].
[6] [Brouwer, 1970].

To this it is possible to object that, if it is true that axiomatic systems set limitations to the activities of mathematicians, it is rather hard to take on board the idea that these limitations are arbitrary and unacceptable.

Such limitations are not arbitrary, because, given an axiomatic system, mathematicians take very seriously questions which concern the independence of the axioms, and whether the axioms are self-evidently true, *ad hoc*, fruitful, etc. And, when it happens that the axiomatizations of mathematical theories are not unique, mathematicians often agonize over which, of the equivalent axiomatizations of a theory, is the most *natural*, elegant, etc.

Furthermore, it seems unreasonable to say that the limitations imposed by an axiom system on mathematical creativity are unacceptable. For, the possibility of changing an axiom system: by extending it through the addition of supplementary axioms, or by rejecting some axioms of the system, or by ditching the whole system and introducing a different one; is rather ample.

Secondly, according to Brouwer, the axiomatic method must be rejected, because it leads mathematicians astray by giving too much importance to structures and their rôle in mathematics.

It seems to me that this, of Brouwer's objections, could receive what would be considered a 'head on' refutation by anyone prepared to accept history of mathematics as the tribunal before which opinions about what is the object of mathematical investigations and the success of such investigations must come. As a matter of fact, XXth century mathematics has become, to a large extent, a science of structure, and this turn taken by mathematical research has proved to be immensely successful also in relation to problems of non-structuralist mathematics.

But, apart from Brouwer's attack against the use of the axiomatic method, there is another very important source of perplexities concerning axiomatic thinking in mathematics. Such a source is to be found in the development of the theory of formal systems.

In the first half of the past century the advances made by the axiomatic method in mathematics achieved a very important result: formal systems themselves became explicit and recognized objects of mathematical investigation. And, as is well known, it did not take long before a number of surprising results were proved concerning the so-called 'limitations of formal systems', surprising results among which Gödel's incompleteness theorems occupy a central position.

The first of these theorems, in the formulation known as the Gödel-Rosser theorem, asserts that any consistent first-order system of arithmetic **A**, whose set of axioms is recursive, is incomplete.

This result reveals two important limitations of formal systems like **A**: (1) they are unable to meet one of the intuitive adequacy conditions for axiomatic systems: that from them it should be possible to prove all the true wffs of their languages; and that (2) there are well formulated mathematical problems of **A** which are not decidable in **A**.

The second incompleteness theorem says that, given a formal system like **A**, it is impossible to prove in **A**, or by means of methods which are representable within **A**, that **A** is consistent.

This last result is very important, because it shows that Hilbert's programme cannot be realized, and that another adequacy condition for formal systems, i.e., consistency, cannot be proved to obtain for **A**, and systems like **A**, unless we use methods of proof whose soundness is at least as dubious as that of those used in **A**.

Two other results are often mentioned in connection with the limitations of formal systems. The first is Church's theorem according to which the validity of wffs belonging to the language of first-order logic is recursively undecidable,[7] and the second is Tarski's theorem which essentially says that the notion of arithmetical truth is not arithmetically definable.

Although Church's theorem shows that even in a complete formal system **F** there might be recursively undecidable questions, it seems to me that such a result cuts both ways, in the sense that it also provides strong evidence in favour of the idea that appealing to axioms is indispensable for the determination of the validity of a wff of **F**.

In fact, if we consider propositional logic (**PL**), we realize that in **PL** there exists an algorithm such that, given any wff. φ belonging to the language $\mathcal{L}_{\mathbf{PL}}$, is able to decide whether or not φ is a tautology. In actual fact, there are several such algorithms (truth-tables, semantic tableaux). Therefore, in **PL** we do not need to prove φ from the axioms to justify our assertion that φ is a tautology. On the other hand, if we want to determine the validity of a wff.' φ belonging to the language of first-order logic, as a consequence of Church's theorem, we need a proof from the axioms.

---

[7]There is no algorithm which, for any wff. $\varphi \in \mathcal{L}_{\mathbf{PrL}}$, is able to determine whether φ is valid or not.

Lastly, concerning the significance of Tarski's theorem, I must say that this result shows that there are definite limitations for what can be expressed within a particular formal system.

Since, the technical results examined above show that certain intuitive adequacy conditions for axiomatic systems cannot be met by a very large class of axiomatic systems F_s, now the obvious question to ask is whether this compromises in any way the axiomatic method.

We have already seen that Gödel's two incompleteness theorems demolish, together with Hilbert's programme, some intuitive ideas about the conditions of acceptability for formal systems, but this does not seem to undermine the use of the axiomatic method in mathematics. For, even though a particular formal system might be incomplete, this could still provide a foundation for the central results of the theory allowing us to consider the formal system, and what we can prove from it, as a well-defined theory on its own right. This theory T, as a consequence of incompleteness, may then be extended into another consistent theory T_1 in which we can prove what we can prove in T as well as results which are not provable in T, and so on.

Incompleteness, therefore, far from undermining the axiomatic method, provides a compelling explanation of the reason why mathematical knowledge does not always grow in a cumulative way, i.e., through the addition of new theorems to the stock of theorems already proved. As the history of mathematics clearly shows, there are, indeed, times when mathematical knowledge grows in a revolutionary way through changes which affect the very foundations of the theory and in so doing generate another theory. These considerations lead us to assert that a mathematical theory is not conceivable as a *heap of theorems*, but, rather, as a sequence of theories which converges to a complete theory (whose set of axioms is not recursive) as to its limit.

Moreover, the impossibility, for a large class of important formal systems, of obtaining completeness, absolute consistency proofs in Hilbert's sense, recursive decidability and full expressibility, shows that, in contrast with the traditional view on these matters (see §2), the axiomatic method is fully compatible with the use of intuition. Indeed, how do we choose the next axiom to add to those we already have in our system to increase expressibility or to mend new anomalies[8] or to prove a certain result, etc.?

---

[8] By 'anomalies' I mean undesirable results and operations such as: one-to-one correspondence between $\mathbb{N}$ and one of its proper subsets before Cantor; algorithms for the

Well, by intuition, by intuition guided by a number of specific prerequisites. (An interesting example of specific prerequisites which direct the choice of axioms in set theory may be found in [Maddy, 1988a] and [Maddy, 1988b].)

The considerations relating to the interaction between intuition and the process of axiom choice lead us naturally to the discussion of the last objection against the axiomatic method which is going to be examined in this paper.

For some authors, the main defect of the axiomatic method consists in being unequal to the task of representing mathematical knowledge,[9] for example, according to Cellucci, the axiomatic method is simply a very successful strategy of justification, which, as a consequence of being considered as 'the' method to be used in doing mathematics, obscures the other face of the coin: the context of mathematical discovery. For him, it is the method of analysis what sheds light on how we acquire mathematical knowledge. And such a method consists in, given a result we want to prove, finding the correct hypotheses from which to prove it.

According to Cellucci, there is an important difference between the axiomatic method and the method of analysis. The axiomatic method presupposes the existence of unchangeable principles given from the beginning. Furthermore, when applied to a mathematical theory, the axiomatic method generates conceptual systems for which there is no exchange of information with the environment (closed systems), because mathematical activity in that theory collapses on to the activity of deriving logical consequences from the axioms. On the contrary, an application of the method of analysis to a particular mathematical theory gives origin to systems for which there is exchange of information with the environment (open systems), because mathematical activity in that theory is not dependent on the existence of unchangeable principles given from the beginning, but on the generation of new hypotheses.

At the beginning of the paper I mentioned some of the important contributions given by the axiomatic method to mathematical heuristics, and I shall not go over these considerations again. But I would like to point out that some of the discoveries obtained through the application of the axiomatic method were extremely useful to bring to the surface objective

---

calculation of the derivative in the analysis of Newton and Leibniz; the Tarski-Banach paradox, etc. etc.

[9][Cellucci, 2000], Ch. VII, §8, p. 254.

limitations of the principles found by means of the method of analysis and used to develop informal theories. A typical example of this comes from set theory. It is well known that Cantor and others tried to prove the Continuum Hypothesis (**CH**) for a long time and without success, and that it was eventually possible to see some light on this matter only when the principles Cantor and others were informally using in developing set theory were patiently individuated, and some of them, opportunely modified so to avoid the known set theoretical paradoxes, were included as axioms in the system **ZFC** by Zermelo and Fraenkel, and Cohen succeeded in producing models of **ZFC** + **CH** and of **ZFC** + ¬**CH** which showed that **CH** is independent of **ZFC**.

Secondly, I do not find any conflict between the axiomatic method and the method of analysis. As a matter of fact, the method of analysis is used all the time in axiomatized theories not only locally, to discover the hypotheses from which to prove a particular result, but also globally in the sense that the method of analysis is used to find the axioms to formalize a given mathematical theory.

Lastly, as we shall see in what follows, I find that the formalization of mathematical theories plays a very important rôle in highlighting how mathematical knowledge grows and, in so doing, gives a very important contribution to mathematical epistemology.

5 The axiomatic method and the quasi-empirical nature of mathematical knowledge

The traditional debate about knowledge operated a basic distinction between knowledge of matters of fact and knowledge of relations of ideas. One of the distinctive features of the traditional debate about knowledge resided in the attempt to provide an answer to the question concerning the nature of mathematical knowledge by means of a clarification of the nature of mathematical judgments/propositions.

Even though there was some difference of opinion among the participants in the debate about how to define predicates like 'analytic', the debate continued unhampered until Quine argued that no demarcation line can be drawn between analytic and synthetic propositions.

The next step was then taken by Lakatos, who decided to shift the centre of attention of the epistemologists from the nature of judgments/propositions to the study of the organization and development of mathematical theories. Work deriving from such a radical change of focus has shown that mathematical knowledge does not always grow in a cumulative way, and that what we call 'Euclidean geometry', 'mathematical analysis', 'set theory', etc. are mathematical research programmes (MRPs). These mathematical research programmes are finite sequences of theories in which an element of the sequence is a *better* theory than its immediate predecessor.

The considerations above are very important for mathematical epistemology, because they show that mathematical theories are quasi-empirical, that is, that the elements of mathematical research programmes contain a large body of revisable statements, even if the theorems produced within them are not:[10]

> ... spatio-temporally singular basic statements whose truth-values are decided by the time-honoured but unwritten code of the experimental scientist.

But, what has the axiomatic method to do with discussions concerning the quasi-empirical nature of mathematical knowledge? The following remarks concerning Cantor-Zermelo set theory will address this question giving an example of how the axiomatic method is able to reveal the quasi-empirical nature of mathematical knowledge.

Cantor-Zermelo set theory historically began with Cantor's informal system known as naïve set theory (**NST**), and in spite of the large number of very interesting and surprising developments undergone by the theory in a relatively short time, at the end of Cantor's activity the system **NST** suffered from a large number of serious problems. On the one hand, results crucial for the development of the theory were still unproved, some such being the well-ordering theorem, the trichotomy of cardinals, the aleph-theorem,[11] and the Continuum Hypothesis. And, on the other hand, **NST**

---

[10][Lakatos, 1967], §2, p. 29.
[11]

Theorem 5.1 (Well-ordering). Every set can be well-ordered.

Theorem 5.2 (Trichotomy of cardinals). For any cardinal numbers α and β, we have that
$$\alpha < \beta \text{ or } \alpha = \beta \text{ or } \alpha > \beta.$$

Theorem 5.3 (Aleph-theorem). Every transfinite cardinal is an aleph.

had been beset by a number of anomalies: the Burali-Forti paradox, Cantor's paradox, Russell's paradox, etc.

The situation changed when Zermelo proved the well-ordering theorem and, sometime after, produced the axiomatic system **Z**. The formulation of this axiomatic system had two main aims: showing that his proof of the well-ordering theorem was sound, and avoiding the paradoxes which affected **NST**.

It is particularly significant that the proof of the well-ordering theorem, and those of the trichotomy of cardinals and of the aleph-theorem, were eventually obtained using the Axiom of Choice (**AC**), because this is a principle of set existence — provably equivalent to the well-ordering theorem — which is not available in **NST**.

In **Z** Zermelo assumes the existence of a domain/universe of objects $\mathcal{B}$ among which are sets, and conceives the membership relation '∈' as holding only among some of the objects contained in $\mathcal{B}$. According to Zermelo, $\mathcal{B}$ is *not* a set and no mathematics can be done with it. Such a position on the universe of sets is particularly important, because it is what provides support for the Zermelian doctrine of *limitation of size*.

According to this doctrine:[12]

> *A collection of sets is a set if and only if it is not equinumerous to the collection of all sets.*

What motivates the doctrine of limitation of size is the idea that the blame for the paradoxes affecting **NST** must be put squarely on the attempt to do mathematics with collections of sets which are 'too large', where a collection of sets A is too large if and only if A is equinumerous to $\mathcal{B}$.

One of the consequences of the implementation of the limitation of size doctrine in **Z** is that the meaning of the word 'set' in **Z** differs from the meaning of the word 'set' in **NST**. Therefore, when **Z** takes over from **NST**, one of the things that are eliminated is the **NST** definition of set according to which[13]

> By an "aggregate" (*Menge*) we are to understand any collection into a whole (*Zusammenfassung zu einem Ganzen*) M of definite and separate objects m of our intuition or our thought. These objects are called the "elements" of M.

---

[12][Fraenkel *et al.*, 1984], Chapter II, §5.3, p. 95.
[13][Cantor, 1895–97], §1, p. 85.

This is an extremely important event also from a philosophical point of view, because it shows that a *Gestalt* shift takes place in going from **NST** to **Z**. This is a *Gestalt* shift which points out that **NST** and **Z** express very different views about sets from one another, and that the growth in mathematical knowledge brought about by the substitution of **Z** for **NST** is not the outcome of a cumulative process.

In fact, much of the content of **NST** is refuted, namely that part of the content of **NST** leading to the contradictions, and the difference between **Z** and **NST** — expressed by new principles of set existence and safe guards against the contradictions — is made particularly sharp and explicit by the axioms of **Z**.

Now, we can also say that **NST** itself is *refuted* by **Z** in the sense that **NST** is superseded by **Z**, which is a better theory than **NST**. And **Z** is a *better theory* than **NST**, because: (i) what you can prove in **NST** without using methods leading to known paradoxes is also provable in **Z**, (ii) there are many results provable in **Z** which are not part of the unrefuted content of **NST**, (iii) the known paradoxes are not derivable in **Z**, and (iv) some of the axioms of **Z**, notably the axiom of choice, have very important applications in various mathematical theories. But, in spite of all this, at some point also **Z** became object of criticism. I shall here consider three points.

First, the concept of *definit* propositional function used in **Z** to prevent the formation of Richard's like paradoxes is vague, because:[14]

> [Zermelo's] definition of "definit" property ... invokes "the universally valid laws of logic"; [and] since Zermelo pays no attention at all to the underlying logic, these laws are left unspecified, and the notion of definite property remains hazy.

Secondly, the set-construction principles to be found in some of the axioms of **Z** are not strong enough to enable us to prove the existence of either the set

$$\{\mathbb{N}, \mathcal{P}(\mathbb{N}), \mathcal{P}(\mathcal{P}(\mathbb{N})), \ldots\}$$

or of the union of its elements. Therefore, in **Z** we cannot prove the existence of sets which have cardinality α, for $\aleph_\omega \leq \alpha$.

Thirdly, as Zermelo himself admits in talking about **Z**, '... the possibility that $x \in x$ is not in itself excluded by our axioms';[15] and, as we know, the

---
[14][van Heijenoort, 1967], p. 199.
[15][Zermelo, 1908a], ibid., p. 203.

condition $x \in x$ is one of the ingredients to be found in the noxious mixture at the root of several set-theoretic paradoxes.

As a response to the above criticism of **Z**, mathematicians have, over a period of time characterized by much controversy, devised the system of set theory known as **ZFC** which, holding on to the basic principles of the Cantor-Zermelo MRP, operates some changes to the axiomatic basis of **Z** which produce a theory, **ZFC**, which represents a marked improvement upon **Z**.

To see this, consider, first of all, that the logic underlining the language of **ZFC** is first-order logic. This choice gives a sharp delineation of what we take the 'universally valid laws of logic' to be, and in so doing answers the first objection raised against **Z** concerning the vagueness of the notion of *definit* propositional function (or property).

Secondly, the introduction of the Axiom Schema of Replacement supplies a principle of set existence able to show that

$$\{\mathbb{N}, \mathcal{P}(\mathbb{N}), \mathcal{P}(\mathcal{P}(\mathbb{N})), \ldots\}$$

is a set.

Thirdly, as a consequence of the Axiom of Foundation, we cannot have situations like the following:

$$(1) \quad A \in A,$$

or like:

$$(2) \quad A \in B_n \in B_{n-1} \in \cdots \in B_1 \in A,$$

where $A, B_1, B_2, \ldots, B_n$ are sets. Likewise if $A, B_1, \ldots$ are sets, we cannot have

$$(3) \quad \cdots \in B_n \in \cdots \in B_1 \in A.$$

Therefore, in **ZFC** a further source of trouble still present in **Z** is removed through the introduction of the axiom of foundation. Another important thing that we must notice is that in **ZFC**, for a collection of sets A to be a set, A, besides not being too large, must also be well-founded, i.e., there must be an $\alpha \in \text{On}$ such that $A \in \mathcal{L}_\alpha$, where $\mathcal{L}_\alpha$ is a level of the cumulative hierarchy $\mathcal{B}$.

The change of meaning undergone by the word 'set' in passing from **Z** to **ZFC**, together with considerations relating to what can be proved in **ZFC** with respect to **Z** and to the new safeguards present in **ZFC** against the paradoxes, etc. show that **Z** and **ZFC** are two different theories, that **Z** is *refuted* by **ZFC**, and that, therefore, the growth of set-theoretical knowledge within the Cantor-Zermelo MRP is not always cumulative.

Here, once more, the axiomatization of **Z** and **ZFC** allows us to draw very sharply and explicitly characteristics and prove limitations of both theories, characteristics and limitations which are at the root of the possibility of making a rational comparison between them.

It is, therefore, the axiomatization of **Z** and **ZFC** what reveals and justifies us in saying that what is known as Cantor-Zermelo set theory is a MRP expressible as a sequence of three quasi-empirical theories, which are given in the following order: **NST, Z, ZFC**.

On the basis of what has been said so far, let us try to address very briefly the main question of the paper: do we really need axioms in mathematics? Well, if by 'really needing axioms in mathematics' we mean that axioms are necessary conditions for mathematical activity to take place, then, given the large number of familiar counterexamples which can be found to this thesis studying the history of mathematics, we can, without hesitation, answer the main question of the paper in the negative.

But, on the other hand, if by 'really needing axioms in mathematics' we mean that the axiomatic method is necessary for the advancement of mathematics, then, if what has been argued in this paper is correct, we are justified in concluding that we really need axioms in mathematics.

BIBLIOGRAPHY

[Brouwer, 1970] L. E. J. Brouwer. *On the foundations of mathematics. Dissertation*, in [Brouwer, 1975], pp. 11–101, 1970.

[Brouwer, 1975] L. E. J. Brouwer. *L. E. J. Brouwer, Collected Works I*, A. Heyting (ed.), North-Holland, Amsterdam, 1975.

[Brouwer, 1990] L. E. J. Brouwer. 'Brouwer Manuscript 3A', in: [van Stigt, 1990].

[Cantor, 1895–97] G. Cantor. *Contributions to the Founding of the theory of Transfinite Numbers*. Dover Publications, Inc., New York, 1955.

[Cellucci, 2000] C. Cellucci. *Le ragioni della logica*, Laterza, Roma-Bari, 2000.

[Fraenkel et al., 1984] A. A. Fraenkel, Y. Bar-Hillel and A. Levy. *Foundations of Set Theory*, Second Revised Edition, North-Holland, Amsterdam, New York, Oxford, 1984.

[Hilbert, 1918] D. Hilbert. 'Axiomatisches Denken', *Math. Ann.*, **vol. 78**, pp. 404–415, 1918. Also in [Hilbert, 1978, pp. 177–188].

[Hilbert, 1978] D. Hilbert. *Ricerche sui Fondamenti della Matematica*, V. M. Abrusci, (ed.), Bibliopolis, Napoli, 1978.

[Lakatos, 1962] I. Lakatos. 'Infinite regress in the foundations of mathematics', in [Lakatos, 1983, pp. 3–23], 1962.

[Lakatos, 1967] I. Lakatos. 'A renaissance of empiricism in the recent philosophy of mathematics', in [Lakatos, 1983, pp, 24–42], vol. II, 1967.

[Lakatos, 1983] I. Lakatos. *Philosophical Papers*, Vol. I and II, Cambridge University Press, Cambridge, 1983.

[Maddy, 1988a] P. Maddy. 'Believing the Axioms I', *The Journal of Symbolic Logic*, **53**, 481–511, 1988.

[Maddy, 1988b] P. Maddy. 'Believing the Axioms II', *The Journal of Symbolic Logic*, **53** 736–764, 1988.

[Plato, 1956] Plato. *Meno*, transl. by W. K. C. Guthrie, Penguin Books, London, 1956.

[van Heijenoort, 1967] J. van Heijenoort, (ed.). *From Frege to Gödel*, Harvard University Press, Cambridge, Massachusetts, 1967.

[van Stigt, 1990] W. P. van Stigt. *Brouwer's Intuitionism*, North-Holland, Amsterdam, 1990.

[Zermelo, 1904] E. Zermelo. 'Proof that every set can be well-ordered', in [van Heijenoort, 1967, pp. 139–141].

[Zermelo, 1908] E. Zermelo. 'A new proof of the possibility of a well-ordering', in [van Heijenoort, 1967, pp. 183–198].

[Zermelo, 1908a] E. Zermelo. 'Investigations in the foundations of set theory I', in [van Heijenoort, 1967, pp. 199–215].

Mathematical Discourse vs. Mathematical Intuition

CARLO CELLUCCI

1 Axiomatic Method and Theology

One of the most uninformative statements one could possibly make about mathematics is that the axiomatic method expresses the real nature of mathematics, i.e., that mathematics consists in the deduction of conclusions from given axioms. For the same could be said about several other subjects, for example, about theology. Think of the first part of Spinoza's *Ethica ordine geometrico demonstrata* or of Gödel's proof of the existence of God, which are both fine specimens of *Theologia ordine geometrico demonstrata*.

To the objection, 'Surely theological entities are not mathematical objects', one could answer: How do you know? If mathematics consists in the deduction of conclusions from given axioms, then mathematical objects are given by the axioms. So, if theological entities satisfy the axioms, why should not they be considered mathematical objects?

Hilbert says: "If in speaking of my points", lines and planes "I think of some system of things, e.g. the system: love, law, chimney sweep ... and then assume all my axioms as relations between these things, then my propositions, e.g. Pythagoras' theorem, are also valid for these things".[1] Similarly he might have said: If in speaking of my points, lines and planes, I think of a suitable triad of theological entities, and assume all my axioms as relations between these things, then my propositions, e.g. Pythagoras' theorem, are also valid for these things.

Indeed, if mathematics consists in the deduction of conclusions from given axioms, then it has no specific content. So it is simply impossible to distinguish geometrical objects, such as 'points, lines and planes', from 'love, law, chimney sweep', or a suitable triad of theological entities. This is vividly

---
[1] Hilbert [1980, p. 40].

illustrated by Russell's statement that "mathematics may be defined as the subject in which we never know what we are talking about, nor whether what we are saying is true".[2]

Hilbert's answer to the objection that, if mathematics consists in the deduction of conclusions from given axioms, then it has no specific content, is that the "circumstance" just "mentioned can never be a defect in a theory, and it is in any case unavoidable".[3] Moreover, "it takes a very large amount of ill will to want to apply" the axioms of geometry to other things "than the ones for which they were meant".[4] For applying them "always requires a certain measure of good will and tactfulness".[5]

Such an answer, however, is inadequate, because appealing to good will and tactfulness is not part of the axiomatic method, and in any case does not provide any specific content for mathematics.

Another answer to the objection that, if mathematics consists in the deduction of conclusions from given axioms, then it has no specific content, is provided by Hintikka. He claims that, contrary to Russell's statement, "we can hope to reach a point where we do know what we are talking about in mathematics, in the sense of being able to formulate descriptively complete theories for different mathematical theories".[6] By 'descriptively complete theories' Hintikka means theories whose models "comprise only the intended models. If there is only one intended model (*modulo* isomorphism), descriptive completeness means categoricity".[7]

Examples of such theories are provided by certain second-order theories, i.e., theories based on second-order logic. Of course, second-order logic is incomplete, in the strong sense that there is no consistent recursive set of rules such that every second-order consequence of second-order axioms can be deduced from such axioms by means of such rules. But, according to Hintikka, this is not in conflict with the view that mathematics consists in the deduction of conclusions from given axioms. The incompleteness of second-order logic does not "force us to search for new axioms, for the old ones", being descriptively complete, "already imply everything. What is needed are stronger and stronger formal rules of logical inference, calculated

---

[2] Russell [1994, p. 76].
[3] Hilbert [1980, p. 41].
[4] *Ibid.*
[5] *Ibid.*
[6] Hintikka [2000, p. 44].
[7] Hintikka [1996, p. 91].

to capture more and more of the model-theoretical consequence relations".[8] Thus "the true novelties are better logical proof methods, not new axioms. In a sense, this would vindicate the idea of mathematics as being concerned primarily with proving theorems from axioms", except that "the proof would not rely exclusively on a closed list of rules of inference but might involve the discovery of new valid rules of inference".[9]

Such an answer, however, is inadequate. As Hintikka himself stresses, in the axiomatic method axioms "are supposed to tell you everything there is to be told" about a given subject matter, and "the rest of your work will consist in merely teasing out the logical consequences of the axioms. You do not any longer need any new observations, experiments or other inputs from reality. It suffices to study the axioms".[10] Now, since second-order logic is incomplete, proving a theorem from given second-order axioms might involve the discovery of new valid rules of inference, which in turn might involve the discovery of new axioms. So proving a theorem from second-order axioms might involve the discovery of new axioms. Then the given second-order axioms do not suffice, you need new observations, experiments or other inputs from reality, external to the given second-order axioms, to discover new axioms. That does not fit in with the axiomatic method.

Even Hintikka acknowledges that, "in practice, such stronger aids of deduction" can "often be codified in the form of new axioms for the mathematical theory in question", so the task of finding such stronger aids of deduction "is not entirely unlike the task of finding stronger and stronger axioms of set theory".[11] But this is hardly compatible with his claim that the fact that the true novelties are better logical proof methods would vindicate the idea of mathematics as being concerned primarily with proving theorems from given axioms. Hintikka goes as far as saying that, "contrary to the oversimplified picture that most philosophers have of mathematical practice, much of what a mathematician actually does is not to derive theorems from axioms".[12] Quite so.

---

[8] Hintikka [2000, p. 44].
[9] *Ibid.*
[10] Hintikka [1996, p. 1].
[11] *Ibid., p. 99.*
[12] *Ibid.,* p. 95.

2   Axiomatic Method and Intuition

Even granting, for argument's sake, that mathematics consists in the deduction of conclusions from given axioms, the question arises: How are axioms justified?

The view that mathematics consists in the deduction of conclusions from given axioms is incapable of providing a satisfactory answer to such crucial question. For axioms, being primary, cannot be proved by discourse, and so any justification of them must appeal to some non-discursive source of knowledge. Now, traditionally, by 'non-discursive source of knowledge', some kind of intuition is understood. But all known justifications of axioms in terms of intuition are inadequate.

This applies in particular to the two main such justifications, which are due to Gödel and Hilbert, respectively.

Gödel claims that axioms are justified if they are based on mathematical intuition. By that he means intellectual intuition — what Kant, though denying that this kind of intuition is humanly possible, calls "pure intuition, intellectual and exempt from the laws of the senses".[13] According to Gödel, "mathematical objects and facts", specifically those of set theory, "exist objectively and independently of our mental acts and decisions".[14] But "they do not belong to the physical world, and even their indirect connection with physical experience is very loose".[15] Still, "despite their remoteness from sense experience, we do have something like a perception also of the objects of set theory".[16] There is no "reason why we should have less confidence in this kind of perception, i.e. in mathematical intuition, than in sense perception, which induces us to build up physical theories".[17] For mathematical intuition "is sufficiently clear to produce the axioms of set theory and an open series of extensions of them".[18]

But how do we get a mathematical intuition of sets sufficiently clear to produce the axioms of set theory? According to Gödel, we may "extend our knowledge of these abstract concepts", that is, "make these concepts themselves precise" and "gain comprehensive and secure insight into the fundamental relations that subsist among them, i.e., the axioms that hold

---

[13] Kant [1900–, II. p. 413].
[14] Gödel [1986–, III. p. 311].
[15] *Ibid.*, II, p. 267.
[16] *Ibid.*, II, p. 268.
[17] *Ibid.*
[18] *Ibid.*

for them", not "by trying to give explicit definitions for concepts and proofs for axioms" — otherwise "one would have nothing from which one could define or prove" — but "rather by cultivating (deepening) knowledge of the abstract concepts themselves".[19] The procedure must consist "in focusing more sharply on the concepts concerned by directing our attention in a certain way, namely, onto our own acts in the use of these concepts, onto our powers in carrying out our acts, etc.".[20] This will "produce in us a new state of consciousness in which we describe in detail the basic concepts we use in our thought, or grasp other basic concepts hitherto unknown to us".[21] Thus we will get an "intuitive grasping of ever newer axioms".[22]

But saying that axioms are justified if they are based on mathematical intuition is inadequate. For suppose that, by Gödel's procedure of focusing more sharply on the concepts concerned, you get an intuition of the concept of set, say Σ. Let S be a formal system for set theory whose axioms such an intuition ensures you to be true of Σ. Since Σ is a model of S, obviously S is consistent. Then, by Gödel's first incompleteness theorem, there is a sentence G of S true of Σ but unprovable in S. Thus $S \cup \{\neg G\}$ is consistent, and hence has a model, say Σ'. Then Σ and Σ' are both models of S, but G is true of Σ and false of Σ', and so Σ and Σ' are not equivalent. Now if, again by Gödel's procedure, you focus more sharply on the way you obtained Σ', you get an intuition of the concept of set Σ'. Then you have two different intuitions, one ensuring that Σ is the genuine concept of set, and the other ensuring that Σ' is the genuine concept of set, where the sentence G is true of Σ and false of Σ'. Which of Σ and Σ' is the genuine concept of set? Gödel's procedure provides no answer.

On the other hand, Hilbert claims that axioms are justified if their consistency can be established by means of a proof based on intuition. By that he means pure sensible intuition — what Kant, though with a different meaning, calls "intuition sensible but pure".[23] According to Hilbert, axioms must be complete, i.e., all mathematical truths of the field concerned must be derivable "from the axioms by means of a finite number of logical inferences".[24] Moreover, axioms must be consistent, in the sense that "a

---

[19] *Ibid.*, III, p. 383.
[20] *Ibid.*
[21] *Ibid.*
[22] *Ibid.*, III, p. 385.
[23] Kant [1900–, II. p. 410].
[24] Hilbert [1996a, p. 1095].

finite number of logical steps based upon them can never lead to contradictory results".[25] The importance of consistency derives from the fact that, "if the arbitrarily given axioms do not contradict one another with all their consequences, then they are true".[26] Thus "'consistent' is identical to 'true'".[27]

But, according to Hilbert, "we can never be certain in advance of the consistency of our axioms if we do not have a special proof of it".[28] To be indisputable, such a special proof must use absolutely reliable methods, and the latter are those based on pure sensible intuition, which is "a kind of intuitive insight".[29] Thus "the most general and fundamental idea of the Kantian epistemology retains its significance: to ascertain the *a priori* intuitive mode of thought".[30] By that, however, Hilbert means something very different from Kant's construction of mathematical concepts. For he claims that "something is already given to us in advance in our faculty of representation: certain extra-logical concrete objects that exist intuitively as an immediate experience before all thought".[31] These objects are "completely surveyable in all their parts, and their presentation, their differences, their succeeding one another or their being arrayed next to one another is immediately and intuitively given to us, along with the objects".[32] Thus Hilbert's view of pure sensible intuition is very un-Kantian.

But saying that axioms are justified if their consistency can be established by means of a proof based on pure sensible intuition, is inadequate. For, by Gödel's second incompleteness theorem, for any formal system containing elementary number theory, no consistency proof based on pure sensible intuition is possible. Hilbert needn't have waited for Gödel to realize that. For Kant had already stressed that "it is, to be sure, a necessary logical condition" that a given concept "must contain no contradiction; but this is not by any means sufficient to guarantee the objective reality of the concept, that is, the possibility of such an object as is thought through the concept".[33]

---

[25] Hilbert [2000, p. 250].
[26] Hilbert [1980, p. 39].
[27] Hilbert [1931, p. 122].
[28] Hilbert [1996c, p. 1120].
[29] Hilbert [1996e, p. 1161].
[30] Hilbert [1996d, pp. 1149–1150].
[31] *Ibid.*, p. 1150.
[32] *Ibid.*
[33] Kant [1900–, p. 187].

Moreover, saying that axioms must be complete is inadequate. For, by Gödel's first incompleteness theorem, every consistent formal system containing elementary number theory is incomplete. Again, Hilbert needn't have waited for Gödel to realize that. For Kant had already stressed that the mathematician "arrives at an illuminating and at the same time general solution of the problem through a chain of inferences that is always guided by intuition".[34] Demonstrations, "as the term itself indicates, proceed through the intuition of the object".[35] An "apodictic proof can be called a demonstration only insofar as it is intuitive".[36] Thus, in the derivation of geometrical theorems we always need new geometrical intuitions, and therefore a purely logical derivation from a finite number of axioms is impossible. Kant's view is shared by Gödel, with a difference. According to Gödel, the assertion that "in the derivation of geometrical theorems we always need new geometrical intuitions, and that therefore a purely logical derivation from a finite number of axioms is impossible", is "incorrect if taken literally".[37] But "if in this proposition we replace the term 'geometrical' by 'mathematical' or 'set-theoretical', then it becomes a demonstrably true proposition".[38]

Furthermore, saying that 'consistent' is identical to 'true' is inadequate. For, by a corollary to Gödel's first incompleteness theorem, every consistent formal system containing elementary number theory has a consistent extension in which some falsity is provable. So consistency is not sufficient for truth. Again, Hilbert needn't have waited for Gödel to realize that. For Kant had already stressed that "a judgment, even though it is free of any internal contradiction, can still be either false or groundless".[39]

It is no escape from Gödel's incompleteness results to claim that, although we must give up "introducing the idea of a total system for mathematics", it is nonetheless possible to consider "the actually existing system of analysis and set theory as providing an adequate framework for accommodating the geometrical and physical disciplines. A formalism may correspond to this aim even without having the property of full deductive closure".[40] This is

---

[34] *Ibid.*, III, p. 471.
[35] *Ibid.*, III, p. 482.
[36] *Ibid.*, III, p. 481.
[37] Gödel [1986–, III. p. 385].
[38] *Ibid.*
[39] Kant [1900–, III. p. 141].
[40] Hilbert-Bernays [1968–70, II, p. 289].

no escape because, even allowing that the actually existing system of analysis and set theory provides such an adequate framework, still, by Gödel's second incompleteness theorem, no consistency proof based on pure sensible intuition is possible for such a system. So the axioms of the system are unjustified.

Similarly, it is no escape from Gödel's incompleteness results to claim that, although no single consistent formal system containing elementary number theory can be complete, we may "avoid as far as possible the effects of Gödel's theorem" if we start from a given incomplete system and "obtain a more complete one by the adjunction as axioms of formulae, seen intuitively to be correct, but which the Gödel theorem shows are unprovable in the original system"; then from this we obtain "a yet more complete system by a repetition of the process, and so on".[41] This process will be continued into the transfinite, associating a system "with any constructive ordinal".[42] This is no escape because the resulting sequence of formal systems will be incomplete.

To justify the axioms of set theory, Gödel and Hilbert appeal to different kinds of intuition, and in different ways.

Gödel thinks it possible to justify the axioms of set theory directly in terms of pure intellectual intuition, since the latter is pure intuition of abstract objects, and sets are abstract objects. In Gödel's view, it is by pure intellectual intuition that "the axioms force themselves upon us as being true".[43]

On the other hand, Hilbert thinks it possible to justify the axioms of set theory only indirectly in terms of pure sensible intuition — by a consistency proof — since pure sensible intuition is pure intuition of concrete objects. A justification of the axioms of set theory can only be indirect since sets are abstract objects. So no direct justification in terms of pure sensible intuition is possible. In Hilbert's view, a consistency proof based on pure sensible intuition "provides us with a justification for the introduction" of propositions concerning abstract objects such as the axioms of set theory, which Hilbert calls the "ideal propositions".[44]

---

[41] Turing [1939, p. 198].
[42] *Ibid.*, p. 161.
[43] Gödel [1986–, II, p. 268].
[44] Hilbert [1967, p. 471].

Gödel's and Hilbert's appeals to intuition can be viewed in a perspective which makes the above comparison between the axiomatic method and theology less haphazard than it might appear.

For, appealing to pure intellectual intuition to justify the axioms of set theory directly, Gödel assigns pure intellectual intuition a power comparable to that Aquinas assigns to God. Aquinas claims that God has the entire knowledge of a thing "by understanding the simple essence".[45] He knows it "with the knowledge of vision", and "the present intuition of God extends over all time, and to all things which exist in any time, as to objects present to him".[46] Similarly Gödel claims that, focusing more sharply on the concepts concerned, we know sets by understanding the simple essence. We know them with the knowledge of vision, and our intuition extends to all sets as to objects present to us. As we have seen, this claim is refuted by Gödel's first incompleteness theorem.

On the other hand, appealing to pure sensible intuition to justify the axioms of set theory indirectly, Hilbert assigns consistency proofs based on pure sensible intuition a role similar to that Spinoza assigns to the ideas of God. Spinoza claims that, "if an architect conceives a building properly constructed, though such a building never existed, and also will never exist, nevertheless the idea of such a building is true; and the idea remains the same, whether the building exists or not".[47] In other words, coherent ideas of the mind are true. For "our mind, in so far as it perceives things truly, is part of the infinite intellect of God".[48] So coherent "ideas of the mind are as necessarily true as the ideas of God".[49] Similarly, Hilbert claims that arbitrarily given axioms which are proved to be consistent by means of a proof based on pure sensible intuition, are true. For consistent ideas of the mind are true. As we have seen, this claim is refuted by a corollary to Gödel's first incompleteness theorem.

Since all known justifications of axioms in terms of intuition are inadequate, we may conclude that the view that the axiomatic method expresses the real nature of mathematics is incapable of providing a satisfactory answer to the question how axioms are justified.

---

[45] Thomas Aquinas, *Summa Theologiae*, I, q. 85, a.5.
[46] *Ibid.*, I, q. 14, a. 9.
[47] Spinoza [1925, II, p. 26].
[48] *Ibid.*, II, p. 125.
[49] *Ibid.*

Against such conclusion it might be objected that it rests on the unproven assumption that any justification of axioms must appeal to some kind of intuition. On the contrary, there is a justification of axioms which does not appeal to intuition: axioms are justified if true consequences follow from them. Therefore it is the consequences that give the reasons why we believe the axioms.

Such an alternative justification of axioms in terms of their consequences has been suggested by several people.

For example, Zermelo claims that "principles must be judged from the point of view of science", i.e., from the point of view of their consequences, "and not science from the point of view of principles fixed once and for all".[50] In particular, the principle of choice is justified because there is "a number of elementary and fundamental theorems and problems" that "could not be dealt with at all without the principle of choice".[51] So long as it "cannot be definitely refuted, no one has the right to prevent the representatives of productive science from continuing to use this 'hypothesis' ".[52]

Similarly, Gödel claims that, "even disregarding the intrinsic necessity of some new axiom, and even in case it has no intrinsic necessity at all, a probable decision about its truth is possible also in another way, namely, inductively by studying its 'success' ", where success means "fruitfulness in consequences, in particular in 'verifiable' consequences".[53] Thus, "besides mathematical intuition, there exists another (though only probable) criterion of the truth of mathematical axioms, namely their fruitfulness in mathematics and, one may add, possibly also in physics".[54]

But a justification of axioms in terms of their consequences is inadequate, because the fact that true consequences follow from the axioms provides no justification for them: true consequences can follow from false axioms.

As Kant states, "inferring the truth of a cognition from the truth of its consequences would be admissible only if all its possible consequences are true", but "this is an unfeasible procedure, since to discern all possible consequences of any accepted proposition exceeds our powers".[55]

---

[50] Zermelo [1967, p. 189].
[51] *Ibid.*, p. 188.
[52] *Ibid.*, p. 189.
[53] Gödel [1986–, p. 261].
[54] *Ibid.*, II, p. 269.
[55] Kant [1900–, III, p. 514].

Evidence for Kant's statement is provided, for example, by the fact that the set of all consequences of the axioms of second-order Peano arithmetic PA^2 is not algorithmically enumerable. For, since the axioms of PA^2 are categorical, a second-order sentence is a consequence of the axioms of PA^2 if and only if it is true of the natural numbers. Thus, if the set of all consequences of the axioms of PA^2 were algorithmically enumerable, so would be the set of all second-order sentences true of the natural numbers. But, by Tarski's theorem for second-order sentences, the set of all second-order sentences true of the natural numbers is not definable in the set of all natural numbers by any second-order formula, and so *a fortiori* it is not algorithmically enumerable. Then the set of all consequences of the axioms of PA^2 is not algorithmically enumerable. Thus there exists no algorithmic procedure, *a fortiori* no feasible procedure, for enumerating all consequences of the axioms of PA^2. Therefore, as Kant states, inferring the truth of a cognition from the truth of its consequences is an unfeasible procedure.

Moreover, a justification of the axioms in terms of their consequences does not fit in with the view that mathematics consists in the deduction of conclusions from given axioms. For such a view involves that, since axioms are what their consequences depend on, it is the axioms that give the reasons for believing the consequences. In particular, according to Hilbert, it is the consistency of the axioms that gives the reason for believing the consequences. Therefore, when supporters of the view that mathematics consists in the deduction of conclusions from given axioms claim that axioms can be justified in terms of their consequences, they speak rather incoherently.

To claim that it is the consequences that give the reasons for believing the axioms one must drop the view that mathematics consists in the deduction of conclusions from given axioms. Actually, one would be well advised to drop that view anyway, because it "does not correspond to simple observation. If the Pythagorean theorem were found to not follow from postulates, we would again search for a way to alter the postulates until it was true. Euclid's postulates came from the Pythagorean theorem, not the other way".[56]

Since a justification of the axioms in terms of their consequences is inadequate, we may finally conclude that the view that mathematics consists in the deduction of conclusions from given axioms is incapable of providing a satisfactory answer to the question how axioms are justified.

---

[56] Hamming [1980, p. 87].

On the other hand, such a view is also incapable of providing a satisfactory answer to the question how axioms are discovered. For, as Aristotle points out, it is "impossible to say anything" about the principles of each science "on the basis of the proper starting points of the science in question, since the starting points are primary to everything".[57] To explain how principles are discovered one needs a method that, "being fit for inquiring, possesses the path to the principles of all disciplines".[58] Such a method is not provided by the axiomatic method. As we will argue in the next section, it is provided by the analytic method.

3 Analytic Method and Discourse

Since antiquity, the main reason for an axiomatic presentation of mathematics has been didactic, i.e., its efficiency in transmitting knowledge in the form of textbooks. But mathematics cannot be reduced to its didactic presentation. Therefore, the assertion that mathematics consists in the deduction of conclusions from given axioms, rather than expressing the real nature of mathematics, at most expresses the real nature of teaching.

This has been stressed by several people. For example, Zermelo distinguishes mathematics from its didactic presentation arguing that "geometry existed before Euclid's *Elements*, just as arithmetic and set theory did before Peano's *Formulaire*, and both of them will no doubt survive all further attempts to systematize them in such a textbook manner".[59]

Actually, teaching is the end of mathematical research — the end not in the sense that it is the goal towards which the activity of progressing through the steps in this sequence is geared, but only in the sense that it is the last term in a sequence. Therefore, it does not express the real nature of mathematics.[60] To express it we must look elsewhere.

Specifically, we must look to the analytic method, a rival of the axiomatic method originally used by Hippocrates of Chios to solve certain mathematical problems, such as cube duplication or the quadrature of certain lunes, and explicitly described by Plato, who used it to solve certain philosophical problems, such as the question whether virtue is teachable or whether the soul is immortal.

---

[57] Aristotle, *Topica*, A 2, 101 a 36-b 1.

[58] *Ibid.*, A 2, 101 b 3-4.

[59] Zermelo [1967, p. 189].

[60] For further discussion of the didactic character of axiomatic presentations, see Cellucci [1998, pp. 127–134].

According to Plato, to solve a problem, "on each occasion I assume the hypothesis which I judge to be the strongest, and I lay down as true whatever seems to me to agree with it", while "I put down as not true whatever does not seem to me to agree with it".[61] However, once you had assumed a hypothesis, you wouldn't go on until "you had investigated its consequences, to see whether they agreed or disagreed with one another".[62] Moreover, you would have to give an account of the hypothesis itself. Now, "to give an account of the hypothesis, you would give it in the same way, assuming another hypothesis, whichever among higher hypotheses seemed best, until you came to something" provisionally "sufficient".[63] And so on, ad infinitum.

However, Plato's description of the analytic method provides no indication as to how hypotheses are obtained. A more complete description can be given as follows.

To solve a mathematical problem, we formulates a hypothesis that is a condition sufficient for its solution. The hypothesis is obtained from the problem, and possibly other data, by non-deductive inferences — inductive, analogical, etc..[64] However, the hypothesis must not only be a condition sufficient for the solution of the problem but must also be plausible. That is, it must be compatible with the existing knowledge, in the sense that, comparing the reasons for and the reasons against the hypothesis on the basis of the existing knowledge, the reasons for the hypothesis prevail over those against it. But the hypothesis is in turn a problem that must be solved, and will be solved much in the same way, i.e. formulating another hypothesis that is a condition sufficient for its solution, and is obtained from the previous hypothesis, and possibly other data, by non-deductive inferences. And so on, ad infinitum. Therefore, the solution of a mathematical problem is an essentially infinite process.

Such a description of the analytic method must be supplemented by a number of remarks.

1. The fact that assessing the plausibility of a hypothesis involves comparing the reasons for and the reasons against the hypothesis on the

---

[61] Plato, *Phaedo*, 100 a 3–7.

[62] *Ibid.*, 101 d 4–5.

[63] *Ibid.*, 101 d 5–e 1.

[64] On different kinds of non-deductive inferences by which hypotheses can be obtained, see Cellucci [2002, Part IV].

basis of the existing knowledge, does not mean that the existing knowledge is final. For, in the process of comparing the reasons for and the reasons against the hypothesis, certain long-standing hypotheses, on which the existing knowledge depends, may turn out to be no longer plausible. That depends on the fact that the investigation concerning the plausibility of the new hypothesis may bring such new data to light, or open such new perspectives, that the balance between the reasons for and the reasons against the long-standing hypotheses is reversed: the reasons against end up prevailing over those for them. In that case, it becomes necessary to modify the long-standing hypotheses or even to drop them.

Of course, when confronted with the choice of adopting a new hypothesis or dropping long-standing ones, there is a natural tendency to stick to the latter. But, in the presence of overwhelming evidence, even long standing hypotheses will in the end be modified, or even dropped.

As Kant points out, a person may be reluctant to "get rid of false hypotheses because he has contrived them and they seem so probable to him", just "like a man who has brought a child up with much effort and care and who afterwards does not want it to go away, so as not to lose all his work, effort, and expense".[65] In fact, when a false consequence is found, one "need not let his spirits sink, just as the alchemist always keeps working on the hypothesis of making gold".[66] Admittedly, "if a cognition has a single false consequence, then it is totally false, even though some right consequences can be derived from it".[67] But even then there is no reason to despair. False hypotheses "serve to get true hypotheses fabricated in subsequent ages, for one who is familiar with all possible false paths cannot possibly fail to find the right path at last".[68] Therefore one must try to make the most of false hypotheses.[69]

2. Since comparing the reasons for and the reasons against a hypothesis

---

[65] Kant [1900–, XXIV, p. 224].

[66] *Ibid.*, XXIV, p. 889.

[67] Kant [1998, I, p. 87].

[68] Kant [1900–, XXIV, p. 225].

[69] On Kant's views concerning hypotheses, probability and verisimilitude, see Capozzi [2002, Ch. 15].

generally involves considering the consequences of the hypothesis, in the analytic method it is the consequences that give the reasons for believing the hypotheses rather than the other way round. Thus, while justifying axioms in terms of their consequences does not fit in with the axiomatic method, it fits in with the analytic method.

This is pointed out by Russell, who argues that "we tend to believe the premises because we can see that their consequences are true, instead of believing the consequences because we know the premises to be true".[70] Now, "the inferring of premises from consequences is the essence of induction; thus the method in investigating the principles of mathematics is really an inductive method, and is substantially the same as the method of discovering general laws in any other science".[71] The "usual mathematical method of laying down certain premises and proceeding to deduce their consequences, though it is the right method of exposition", i.e., the right didactic method, "does not, except in the more advanced portions, give the order of knowledge".[72] The actual order of knowledge is, first, "the registration of 'facts'", then "the inductive discovery of hypotheses, or logical premises, to fit the facts", and finally "the deduction of new propositions from the facts and hypotheses".[73] Thus there is a "close analogy" between the methods of pure mathematics and the methods of the sciences of observation".[74]

Russell's version of the analytic method has a definite Aristotelian flavor. For Aristotle claims that "induction is the starting-point which knowledge even of the universal presupposes", since "there are starting-points from which syllogism proceeds, which are not reached by syllogism; it is therefore by induction that they are acquired".[75] Such a version of the analytic method, however, is an essentially incomplete one. For, in the analytic method, hypotheses may be obtained not only by induction but also by other kinds of non-deductive inferences. Moreover, the justification of hypotheses may proceed not only downwards, considering their consequences, but also upwards, formulating new hypotheses.

---

[70] Russell [1973, pp. 273–274].
[71] *Ibid.*, p. 274.
[72] *Ibid.*, p. 282.
[73] *Ibid.*
[74] *Ibid.*, p. 272
[75] Aristotle, *Ethica Nicomachea*, VI 3, 1139 b 28–31.

3. That, in the analytic method, the solution of mathematical problems is an essentially infinite process, does not mean that the process of passing from one hypothesis to another cannot stop temporarily. In fact, it stops temporarily at each step. For, to assess the hypothesis formulated at a given step, one must compare the reasons for and the reasons against it, and that may require a long time, since it may involve considering many consequences of the hypothesis. In any case, the process will always stop only temporarily. Sooner or later it will have to start again, since every hypothesis is a problem, and a problem that must be solved. The fact that, at a given step, the reasons for a hypothesis prevail over those against it, provides only a temporary support for the hypothesis that can be reversed at any time. For their prevailing is only relative to the existing knowledge, and may be upset if new data are brought to light or new perspectives are opened.

4. The analytic method is both a method of discovery and a method of justification. This depends on the fact that a peculiar character of non-deductive inferences is that they allow us to infer different conclusions from the very same premisses. For example, from the premiss, 'All the digits in the decimal expansion of π computed so far are random', one may inductively infer both 'All the digits in the decimal expansion of π are random' and 'All the digits in the decimal expansion of π computed so far are random, but those which will computed in the future will consist of all 9's'. Since non-deductive inferences allow us to infer different conclusions from the very same premisses, to find a suitable hypothesis one must choose between different conclusions. That requires a careful assessment of the reasons for and the reasons against each conclusion. Such an assessment is a process of justification, therefore justification is part of discovery. This blurs the distinction between discovery and justification and makes the analytic method both a method of discovery and a method of justification. Thus the distinction between discovery and justification loses its theoretical importance.

Supporters of the view that the axiomatic method expresses the real nature of mathematics motivate such distinction arguing that discovery escapes logical analysis. According to them, one must distinguish sharply between the process of discovering new mathematical results and the process of justifying mathematical results already discovered.

The former is purely subjective and is left to psychological analysis, the latter is more definite and may be subject to logical analysis.

For example, Frege claims that "it not uncommonly happens that we first discover the content of a proposition, and only later give the rigorous proof of it", therefore in general "the question of how we arrive at the content of a judgement should be kept distinct from the other question, Whence do we derive the justification for its assertion?".[76] The "first question may have to be answered differently for different persons; the second is more definite, and the answer to it is connected with the inner nature of the proposition considered".[77] Therefore the first question is purely psychological, only the second one "is removed from the sphere of psychology".[78]

This, however, overlooks that, if discovery escapes logical analysis, then justification too escapes it. For, as we have seen, the view that the axiomatic method expresses the real nature of mathematics is incapable of providing a logical analysis of the justification of the axioms, and, without such a justification, the axiomatic method does not justify anything.

5. While intuition plays an essential role in the axiomatic method, it plays no role in the analytic method, either in the discovery or in the justification of hypotheses. For a hypothesis for the solution of a given problem is obtained from the problem, and possibly other data, by means of non-deductive inferences, thus not by intuition but by discourse. Moreover, the plausibility of the hypothesis is established comparing the reasons for and the reasons against it, thus not by intuition but by discourse. Therefore, in the analytic method, intuition is replaced by discourse.

This entails that the analytic method cannot be absolutely reliable. For, the conclusions of the non-deductive inferences by which hypotheses are obtained do not follow from their premises with absolute necessity, since non-deductive inferences are ampliative. Moreover, the process by which the plausibility of hypotheses is established is not absolutely reliable. For it consists in comparing the reasons for

---

[76] Frege [1959, p. 3].
[77] Frege [1967, p. 5].
[78] Frege [1959, p. 3].

and the reasons against them, and such a comparison depends on the knowledge existing at that moment, which is not absolutely reliable. Therefore mathematics cannot be absolutely certain.

Supporters of the view that the axiomatic method expresses the real nature of mathematics claim that, being based on intuition, the axiomatic method is absolutely reliable. For, in their opinion, intuition provides an absolutely reliable justification for the axioms, and conclusions of deductive inferences follow from their premises with absolute necessity. Therefore mathematics is absolutely certain. But this claim is unfounded because, as we have already seen, all known justifications of axioms in terms of intuition are inadequate.

6. The analytic method should not be confused with the analytic-synthetic method, originally stated by Aristotle and restated by Pappus and others in terms of a different view of the direction of analysis.[79] Such two methods are essentially different.[80] As Lakatos points out, in the analytic method, analysis is a means of finding unknown hypotheses, and proceeds "without any known lemmas, without any safe axiomatic systems".[81] On the contrary, in the analytic-synthetic method, analysis is simply "a heuristic pattern in already axiomatized Euclidean geometry".[82] As a means of finding hypotheses it loses "its function; when used at all", it is "only a heuristic device for mobilizing the — already proven or trivially valid — lemmas necessary for the synthesis".[83] Analysis is "not any more a venture into the unknown", but only "an exercise in mobilizing and ingeniously connecting the relevant parts of the known. The lemmas which were once daring and often falsified conjectures harden into auxiliary theorems".[84]

While the analytic-synthetic method, being a heuristic pattern in already axiomatized Euclidean geometry which can be used only for finding proofs of given propositions from given axioms, essentially de-

---

[79] On Aristotle's form of the analytic-synthetic method, see Byrne [1997]. On Pappus' variant, see Hintikka-Remes [1974], Knorr [1993].

[80] This is usually overlooked; e.g., see Mueller [1992], Menn [2002]. On the distinction between these two methods, see Cellucci [1998, Ch.8].

[81] Lakatos [1978, II, p. 99].

[82] Ibid., II, p. 100.

[83] Ibid.

[84] Ibid.

pends on the axiomatic method, the analytic method is independent of it and indeed alternative to it. Actually, the axiomatic method is what results from the analytic method if, at a certain stage, the process of passing from one hypothesis to another is stopped definitively, the hypothesis reached at that step being considered no longer as a problem to be solved but as an absolutely unproblematic starting point. Therefore the axiomatic method is an unjustified truncation of the analytic method which removes the most important part of it.

This explains Plato's vehement attack against the axiomatic method. He claims that mathematicians practising the axiomatic method use hypotheses improperly, because they "take them for granted, as axiomatic principles, and do not think it necessary to give any account of them either to themselves or to others, considering them as absolutely evident. Then, starting from these hypotheses and developing their consequences, they arrive at last at the result they were aiming at, and conclude 'What it was required to do' ".[85] Since they do not give any account of the hypotheses, they "only dream about being, and never can they behold the waking reality so long as they leave the hypotheses which they use unexamined, and are unable to give an account of them".[86] But, "when a man does not know his own starting-point, and when the conclusion and intermediate steps are also woven together out of unknown material, how can he imagine that such a fabric of convention can ever become science?"[87]

7. It is interesting to note that supporters of the view that the axiomatic method expresses the real nature of mathematics often make claims that don't fit in with the axiomatic method but only with the analytic method.

An example is provided by Hintikka's claim that proving a theorem from given second-order axioms might involve the discovery of new valid rules of inference, which in turn might involve the discovery of new axioms. Another example is provided by Zermelo's claim that the axiom of choice is justified by its consequences.

An even more significant example is provided by Hilbert's claim that,

---

[85] Plato, *Respublica*, VI 510 c 6-d 3.
[86] *Ibid.*, VII 533 b 8-c 3.
[87] *Ibid.*, VII 533 c 3–5.

in solving mathematical problems, we "do not habitually follow the chain of reasoning back to the axioms in arithmetical discussions, any more than in geometrical".[88] Rather, we proceed "by means of a finite number of steps based upon a finite number of hypotheses which are implied in the statement of the problem and which must be exactly formulated".[89] Such hypotheses are temporarily taken "as the axioms of the individual fields of knowledge".[90] But the solution thus obtained is "only temporary. In fact, in the individual fields of knowledge" the need soon arises "to ground the fundamental axiomatic propositions themselves".[91] Thus one gives 'proofs' of them. However, "critical examination of these 'proofs' shows that they are not in themselves proofs, but basically only make it possible to trace things back to certain deeper propositions, which in turn are now to be regarded as new axioms instead of the propositions to be proved".[92] And so on. Therefore, we solve mathematical problems by the analytic method.

Surprisingly enough, Hilbert even states that the "regressive method", i.e., the analytic method, "finds its perfect expression in what is called today the 'axiomatic method' ".[93] On the other hand, however, he states: "We call the development of a theory axiomatic, when its basic concepts and basic assumptions are put as such at its beginning, and the remaining content of the theory is derived logically from them by means of definitions and proofs. In this sense, geometry was axiomatically founded by Euclid".[94]

Such patently inconsistent statements can be reconciled only if one takes into account that they were made at different stages. Originally, as a working mathematician, Hilbert was naturally led to think that we solve mathematical problems by the analytic method. But his attitude drastically changed following Weyl's and Brouwer's attacks against classical mathematics, which threatened "to dismember and mutilate our science".[95] Then his main worry became "to regain

---

[88] Hilbert [2000, p. 246].
[89] *Ibid.* p. 244.
[90] Hilbert [1996b, p. 1108].
[91] *Ibid.*, p. 1109.
[92] *Ibid.*
[93] Hilbert [1992, p. 18].
[94] Hilbert-Bernays [1968–70, I, p. 1].
[95] Hilbert [1996c, p. 1119].

for mathematics the old reputation for incontestable truth".[96] He thought that the method to achieve that aim was "none other than the axiomatic".[97] Then 'axiomatic method' could no longer designate the analytic method. Unfortunately for Hilbert, however, Gödel's incompleteness results showed that the axiomatic method was unequal to Hilbert's intended aim.

4 Demonstrative and Non-Demonstrative Reasoning

The contrast between the axiomatic method and the analytic method is in particular a contrast between the two kinds of reasoning on which the two methods are based, i.e., demonstrative and non-demonstrative reasoning, respectively. Demonstrative reasoning is the deductive derivation of conclusions from premises which are primitive and true, in some sense of 'true'. Non-demonstrative reasoning is the non-deductive derivation of conclusions from premises which are not known to be true but are only accepted opinions, i.e., plausible propositions.

That non-demonstrative reasoning is the non-deductive derivation of conclusions from premises which are not known to be true originates the objection against the analytic method that, while demonstrative reasoning, on which the axiomatic method is based, is cogent, non-demonstrative reasoning, on which the analytic method is based, is not cogent.

Building on such objection an influential tradition, from Aristotle to Polya, claims that there exists a sharp distinction between demonstrative and non-demonstrative reasoning, and that the former is essentially superior to the latter.

According to such tradition, "there are two kinds of reasoning", i.e., "demonstrative reasoning" and non-demonstrative reasoning — what Aristotle calls 'dialectical' and Polya "plausible reasoning".[98] Demonstrative reasoning "starts from premises that are true and primitive".[99] It draws conclusions from them by syllogism, which is truth-preserving. Therefore demonstrative reasoning is "safe, beyond controversy, and final".[100] Conclusions obtained by means of it are absolutely certain. For that reason,

---

[96] *Ibid.*
[97] *Ibid.*
[98] Polya [1954, I, p. vi].
[99] Aristotle, *Topica*, A 1, 100 a 27.
[100] Polya [1954, I, p. v].

"we secure our mathematical knowledge by demonstrative reasoning".[101] On the other hand, non-demonstrative reasoning starts "from opinions that are accepted", i.e., "shared by everyone, or by most people, or by the wise, and of them, by all, or by most, or by the best known and illustrious".[102] It draws conclusions from them not only by syllogism but also by induction, for "there is on the one hand induction, on the other syllogism".[103] Therefore non-demonstrative reasoning "is hazardous, controversial, and provisional".[104] Conclusions obtained by means of it are not absolutely certain. Admittedly, demonstrative reasoning is "incapable of yielding essentially new knowledge about the world around us".[105] And "anything new that we learn about the world involves plausible reasoning".[106] I.e., it involves non-demonstrative reasoning. In particular, the latter provides "the path to the principles of all sciences".[107] But "the result of the mathematician's creative work is demonstrative reasoning", which "is his profession and the distinctive mark of his science".[108]

The Aristotle–Polya tradition is still so influential that, consciously or unconsciously, most contemporary debate concerning the relations between demonstrative and non-demonstrative reasoning derives from it. However, its claim that there exists a sharp distinction between demonstrative and non-demonstrative reasoning, and that the former is essentially superior to the latter, is untenable. This can be shown as follows.

To know whether an argument is demonstrative, one must know whether its premisses are true. But knowing whether they are true is generally impossible. This is implicit in what we have already said, but there is no harm in repetition.

That the premisses are true can be meant either in Gödel's strong sense that they express properties of objects independent of us, or in Hilbert's weak sense that they are consistent.

If the premisses are true in Gödel's strong sense that they express properties of objects independent of us, then they have a model, which consists of such objects. However, by Gödel's first incompleteness theorem, the propo-

---

[101] *Ibid.*
[102] Aristotle, *Topica*, A 1, 100 a 30, 100 b 21–23.
[103] *Ibid.*, A 12, 105 a 11–12.
[104] Polya [1954, I, p. v.]
[105] *Ibid.*
[106] *Ibid.*
[107] Aristotle, *Topica*, A 2, 101 b 3–4.
[108] Polya [1954, I, p. vi].

sition, 'The premises have a model', is not provable from them but only from a proper extension of them. If the premises of such proper extension are true in Gödel's strong sense, then they have a model. However, by Gödel's first incompleteness theorem, the proposition, 'The premises of the proper extension have a model', is not provable from them but only from a proper extension of them. And so on, ad infinitum. Therefore, it is impossible to know whether the premises are true in Gödel's strong sense.

On the other hand, if the premises are true in Hilbert's weak sense that they are consistent, then, by Gödel's second incompleteness theorem, the consistency of the premises is not provable from them but only from a proper extension of them. If the premises of such proper extension are true in Hilbert's weak sense, then, by Gödel's second incompleteness theorem, the consistency of the premises of the proper extension is not provable from them but only from a proper extension of them. And so on, ad infinitum. Therefore, it is impossible to know whether the premises are true in Hilbert's weak sense.

We may then conclude that knowing whether the premises are true is generally impossible.

This is often overlooked. For example, Dummett claims that, for deductive reasoning "to be fruitful, we must be able to grasp the premises and acknowledge them as true without perceiving the possibility of drawing" the "conclusion".[109] He also claims that deductive reasoning is "astonishingly fruitful".[110] Therefore, Dummett implicitly assumes that we are able to grasp the premises and acknowledge them as true.

Since knowing whether the premises are true is generally impossible, the premises of demonstrative arguments are only accepted opinions. So they have the same status as the premises of non-demonstrative arguments. Thus our reliance on demonstrative reasoning ultimately rests on non-demonstrative reasoning, if not on faith.

This is acknowledged even by Polya, who admits that non-demonstrative reasoning, and specifically "analogy and particular cases", perhaps "not only help to shape the demonstrative argument and to render it more understandable, but also add to our confidence in it. And so we are led to suspect that a good part of our reliance on demonstrative reasoning may come from plausible reasoning".[111]

---

[109] Dummett [1991, p. 305].
[110] Ibid., p. 306.
[111] Polya [1954, II, p. 168].

That it is impossible to know whether the premises are true disposes of the objection against the analytic method that, while demonstrative reasoning is cogent, non-demonstrative reasoning is not cogent. Such an objection is untenable, because demonstrative reasoning cannot be more cogent than the premises from which it starts. But the premises cannot be cogent, since knowing whether they are true is generally impossible. So they are only accepted opinions, and therefore have the same status as the premises of non-demonstrative reasoning. This blurs the distinction between demonstrative and non-demonstrative reasoning.

Moreover, not only it is impossible to know whether the premises are true, bit it is also impossible to justify deductive inferences in any absolute sense. This disposes of the further objection against the analytic method that, while deductive inferences can be justified, non-deductive inferences cannot be justified. Such an objection is untenable, because deductive inferences can be justified as much, or as little, as non-deductive inferences. Indeed, they can be justified much in the same sense as non-deductive inferences, specifically, in a sense that is by no means absolute.[112] In fact, the question of justifying inferences is badly formulated by supporters of the view that the axiomatic method expresses the real nature of mathematics. For they assume that inferences can be justified merely referring to the internal logical structure of the inferences, whereas a justification — though a not absolute one — can be given only in terms of the role inferences play in knowledge. Again, the fact that deductive inferences can be justified much in the same sense as non-deductive inferences, blurs the distinction between demonstrative and non-demonstrative reasoning.

We may then conclude that the claim of the Aristotle-Polya tradition that there exists a sharp distinction between demonstrative and non-demonstrative reasoning, and that the former is essentially superior to the latter, is untenable. It must then be replaced by the claim that there is no sharp difference between demonstrative and non-demonstrative reasoning, and that the former is by no means essentially superior to the latter.

Such a claim has been actually made by another tradition, from Plato to Ramus. According to such tradition, the analytic method based on non-demonstrative reasoning, which such tradition calls 'dialectic', "is the only way which, doing away with hypotheses, is capable of taking to the starting-

---

[112]Space prevents me from discussing this matter here. I will discuss it in Cellucci [200?].

point itself in order to make the conclusions secure".[113] In other words, it is capable of taking us "to the vivid original forms, the archetypes".[114]

There is, however, a feature of the Plato-Ramus tradition that is definitely untenable. According to such tradition, non-demonstrative reasoning is absolutely reliable. For dialectic "shows us how not to lose the right way in arguing".[115] Armed with dialectic, one "can fight all the objections one by one and refute them, not by appeals to opinion, but to absolute truth, never faltering at any step of the argument".[116] One can do that because dialectic is capable of taking to the starting-point itself by means of intellectual intuition. Of all faculties in the soul, "intellectual intuition answers to the highest".[117] Indeed it is able "to behold those higher things, which can only be seen with the mind's eye".[118] Therefore conclusions obtained by means of non-demonstrative reasoning are absolutely certain.

But non-demonstrative reasoning cannot be absolutely reliable, not only because it consists of non-deductive inferences, which, being ampliative, are not absolutely reliable, but also because the process by which the plausibility of its premises is established is not absolutely reliable. Therefore conclusions obtained by means of non-demonstrative reasoning cannot be absolutely certain.

The claim of the Plato–Ramus tradition that they are absolutely certain depends on the assumption that dialectic can "reach the starting-point of everything, which depends on no hypothesis".[119] It can "ascend to the vision of the infinite mind".[120] But such an assumption is untenable because, as we have seen, contrary to what the Plato–Ramus tradition maintains, we cannot ascend to the vision of the infinite mind by Gödel's mathematical intuition, i.e. by intellectual intuition.

That conclusions obtained by non-demonstrative reasoning cannot be absolutely certain does not contradict the above statement that demonstrative reasoning is by no means essentially superior to non-demonstrative reasoning. For the claim of the Aristotle-Polya tradition, that conclusions obtained by means of demonstrative reasoning are absolutely certain, is un-

---
[113] Plato, *Respublica*, VII, 533 c 7-d 1.
[114] Ramus [1964, p. 43 v].
[115] *Ibid.*, p. 8 r.
[116] Plato, *Respublica*, VII, 534 c 1–3.
[117] *Ibid.*, VI, 511 d 8.
[118] *Ibid.*, VI, 511 a 1.
[119] *Ibid.*, VI, 511 b 6–7.
[120] Ramus [1964, p. 43 v].

tenable. Such conclusions cannot be absolutely certain, not only because it is generally impossible to know whether the premisses of demonstrative arguments are true and deductive inference cannot be justified in any absolute sense, but also because it is generally impossible to verify the correctness of demonstrative arguments, even when they are obtained by formal inference rules. For some demonstrative arguments are so long and complex that one can never be sure that they contain no mistakes.

Thus, while the Plato-Ramus tradition mistakenly claims that conclusions obtained by means of non-demonstrative reasoning are absolutely certain, the Aristotle-Polya tradition mistakenly claims that conclusions obtained by means of demonstrative reasoning are absolutely certain.

That conclusions obtained by means of demonstrative or non-demonstrative reasoning cannot be absolutely certain, entails that mathematical knowledge cannot be absolutely certain.

Hume claims that no mathematician places "entire confidence in any truth immediately upon his discovery of it", and although his confidence increases "every time he runs over his proofs" but "still more by the approbation of his friends" and is raised "to its utmost perfection by the universal assent and applauses of the learned world", still "this gradual encrease of assurance is nothing but the addition of new probabilities".[121] Therefore, all mathematical knowledge "resolves itself into probability, and becomes at last of the same nature with that evidence, which we employ in common life".[122]

Hume's claim is a perfectly sensible one, with two changes. First, the approbation of the mathematician's friends and the universal assent and applauses of the learned world — Hume's version of Aristotle's accepted opinions — should be understood as ultimately based on an assessment of the reasons for and the reasons against the premisses from which proofs start. For the correctness of proofs essentially depends on the plausibility of such premisses. Second, 'probability' should be replaced by 'plausibility'. For probability is a quantitative thing, whereas plausibility is not. With these two changes, Hume's conclusion is converted into the conclusion that all mathematical knowledge resolves itself into plausibility, and becomes at last of the same nature with that evidence, which we employ in common life.

---

[121] Hume [1978, p. 180].
[122] *Ibid.*, p. 181.

Against the conclusion that mathematical knowledge cannot be absolutely certain and resolves itself into plausibility it might be objected that it essentially depends on Gödel's incompleteness results. But, if mathematical knowledge cannot be absolutely certain, then Gödel's incompleteness results too are not absolutely certain. Therefore the conclusion that mathematical knowledge cannot be absolutely certain is not absolutely certain.

Such an objection, however, neglects that using Gödel's incompleteness results to conclude that mathematical knowledge cannot be absolutely certain consists in a *reductio ad absurdum*. Let us suppose for argument's sake that mathematical knowledge is absolutely certain. Then Gödel's incompleteness results are absulutely certain. But they entail that mathematical knoweldge cannot be absolutely certain. Contradiction. Therefore mathematical knowledge cannot be absolutely certain. This argument uses *reductio ad absurdum*, but the certainty of *reductio ad absurdum* is part of the assumption that mathematical knoweldge is absolutely certain, for many mathematical arguments essentially depend on *reductio ad absurdum*.

That mathematical knowledge cannot be absolutely certain is inescapable because, "about things that are not perceptible", the "gods alone have a certain knowledge, humans may only form hypotheses".[123] The most they can do is to "adopt the best and least refutable of human hypotheses and, embarking on it as a kind of raft, run the risk of sailing the seas of life".[124] Any serious reflection on the nature of mathematics must start from here.[125]

Acknowledgements

I am very indebted to Reuben Hersh and Robert Thomas for suggestions concerning an earlier draft of this paper.

BIBLIOGRAPHY

[Byrne, 1997] Patrick H. Byrne. *Analysis and Science in Aristotle*. State University of New York Press, Albany, 1997.

[Capozzi, 2002] Mirella Capozzi. *Kant e la logica*, vol. I. Bibliopolis, Napoli, 2002.

[Cellucci, 1998] Carlo Cellucci. *Le ragioni della logica.*, Laterza, Roma-Bari, 1998. Review by Donald Gillies in *Philosophia Mathematica*, **7**, pp. 213–222, 1999.

---

[123] Diels [1964, 24 B 1 (Alcmaeon)].
[124] Plato, *Phaedo*, 85 c 8-d 2.
[125] The views expressed in this paper are strictly related to those of Cellucci [1998; 2000; 2002].

[Cellucci, 2000] Carlo Cellucci. The Growth of Mathematical Knowlege: An Open World View. In Emily R. Grosholz and Herbert Breger (eds.), *The Growth of Mathematical Knowedge*, Kluwer, Dordrecht, pp. 153–176, 2000.

[Cellucci, 2002] Carlo Cellucci. *Filosofia e matematica*. Laterza, Roma-Bari, 2002. Review by Donald Gillies in *Philosophia Mathematica*, **11**, pp. 246–253, 2003. An English revised version of the Introduction will appear in Reuben Hersh (ed.), *18 Unconventional Essays on the Nature of Mathematics*, Springer-Verlag, New York.

[Cellucci, 200?] Carlo Cellucci. *Filosofia e conoscenza*. Laterza, Roma-Bari, in preparation.

[Diels, 1964] Hermann Diels. *Die Fragmente der Vorsokratiker*, Walther Krantz (ed.), Weidman, Berlin, 1964.

[Dummett, 1991] Michael Dummett. *Frege. Philosophy of Mathematics*. Duckworth, London, 1991.

[Ewald, 1996] William Ewald (ed.). *From Kant to Hilbert. A Source Book in the Foundations of Mathematics*. Oxford University Press, Oxford, 1996.

[Frege, 1959] Gottlob Frege. *The Foundations of Arithmetic*. Blackwell, Oxford, 1959.

[Frege, 1967] Gottlob Frege. Begriffsschrift, a Formula Language, modeled upon that of Arithmetic, for Pure Thought. In [van Heijenoort, 1967, pp. 5–82].

[Gödel, 1986–] Kurt Gödel. *Collected Works*. Solomon Feferman et al. (eds.), Oxford University Press, Oxford, 1986–.

[Hamming, 1980] Richard Wesley Hamming. The Unreasonable Effectiveness of Mathematics. *The American Mathematical Monthly*, **87**, pp. 81–90, 1980.

[Hilbert, 1931] David Hilbert. Beweis des Tertium non datur. *Nachrichten von der Gesellschaft der Wissenschaften zu Göttingen*, Mathematisch-Physikalische Klasse, pp. 120–125, 1931.

[Hilbert, 1967] David Hilbert. The Foundations of Mathematics. In [van Heijenoort, 1967, pp. 464–479].

[Hilbert, 1980] David Hilbert. Letter to Frege 29.12.1899. In Gottlob Frege, *Philosophical and Mathematical Correspondence*, Gottfried Gabriel et al. (eds.), The University of Chicago Press, Chicago, pp. 38–41, 1980.

[Hilbert, 1992] David Hilbert. *Natur und mathematisches Erkennen. Vorlesungen, gehalten 1919–1920 in Göttingen*, David E. Rowe (ed.), Birkhäuser, Basel, 1992.

[Hilbert, 1996a] David Hilbert. On the Concept of Number. In [Ewald, 1996, pp. 1092–1095].

[Hilbert, 1996b] David Hilbert. Axiomatic Thought. In [Ewald, 1996, pp. 1107–1115].

[Hilbert, 1996c] David Hilbert. The New Grounding of Mathematics, First Report. In [Ewald, 1996, pp. 1115–1134].

[Hilbert, 1996d] David Hilbert. The Grounding of Elementary Number Theory. In [Ewald, 1996, pp. 1148–1157].

[Hilbert, 1996e] David Hilbert. Logic and the Knowledge of Nature. In [Ewald, 1996, pp. 1157–1165].

[Hilbert, 2000] David Hilbert. Mathematical Problems. In Jeremy J. Gray, *The Hilbert Challenge*, Oxford University Press, Oxford, pp. 240–286, 2000.
[Hilbert and Bernays, 1968–70] David Hilbert and Paul Bernays. *Grundlagen der Mathematik*. Springer-Verlag, Berlin, 1968–70.
[Hintikka, 1996] Jaakko Hintikka. *The Principles of Mathematics Revisited*. Cambridge University Press, Cambridge, 1996.
[Hintikka, 2000] Jaakko Hintikka. *On Gödel*. Wadsworth, Belmont, CA, 2000.
[Hintikka and Remes, 1974] Jaakko Hintikka and Unto Remes. *The Method of Analysis. Its Geometrical Origins and Its General Significance*. Reidel, Dordrecht, 1974.
[Hume, 1978] David Hume. *A Treatise of Human Nature*, Lewis Amherst Selby-Bigge and Peter Harold Nidditch (eds.), Oxford University Press, Oxford, 1978.
[Kant, 1900–] Immanuel Kant. *Gesammelte Schriften*, Königlich Preußischen Akademie der Wissenschaften (ed.), De Gruyter, Berlin, 1900–.
[Kant, 1998] Immanuel Kant. *Logik-Vorlesung. Unveröffentlichte Nachschriften I-II*, Tillmann Pinder (ed.), Meiner, Hamburg, 1998.
[Knorr, 1993] Wilbur Richard Knorr. *The Ancient Tradition of Geometric Problems*, Dover, New York, 1993.
[Lakatos, 1978] Imre Lakatos. *Philosophical Papers*. Cambridge University Press, Cambridge, 1978.
[Menn, 2002] Stephen Menn. Plato and the Method of Analysis. *Phronesis*, **47**, pp. 193–223, 2002.
[Mueller, 1992] Ian Mueller. Mathematical Method, Philosophical Truth. In Richard Kraut (ed.), *The Cambridge Companion to Plato*, Cambridge University Press, Cambridge, pp. 170–199, 1992.
[Polya, 1954] George Polya. *Mathematics and Plausible Reasoning*, Princeton University Press, Princeton, 1954.
[Ramus, 1964] Petrus Ramus (Pierre de la Ramée). *Dialecticae institutiones*. In Petrus Ramus, *Dialecticae institutiones. Aristotelicae animadversiones*, Wilhelm Risse (ed.), Frommann, Stuttgart-Bad Cannstatt, 1964.
[Russell, 1973] Bertrand Russell. *Essays in Analysis*, Douglas Lacey (ed.), Allen & Unwin, London, 1973.
[Russell, 1994] Bertrand Russell. *Mysticism and Logic*. Routledge, London, 1994.
[Spinoza, 1925] Bento Spinoza. *Opera*, Carl Gebhardt (ed.). Winters, Heidelberg, 1925.
[Turing, 1939] Alan Turing. Systems of Logic Based on Ordinals. *Proceedings of the London Mathematical Society*, **45**, pp. 161–228, 1939.
[van Heijenoort, 1967] Jean van Heijenoort, (ed.). *From Frege to Gödel. A Source Book in Mathematical Logic, 1879–1931*. Harvard University Press, Cambridge, Mass., 1967.
[Zermelo, 1967] Ernst Zermelo. A New Proof of the Possibility of a Well-Ordering. In [van Heijenoort, 1976, pp. 183–198].

Mathematical Explanation

MARY LENG

The theme of this book is 'Mathematical Reasoning and Heuristics'. In this paper, I would like just to focus on the first of these topics, mathematical reasoning, and to consider in particular some philosophical issues that arise when we use mathematical reasoning in the natural sciences to explain empirical phenomena. I am particularly interested in the ontological commitments that may be thought to be incurred by those who wish to offer mathematical explanations of empirical phenomena. On this issue, Alan Baker and Mark Colyvan have both recently argued that the appeal to mathematics in the context of scientific explanation puts particular pressure on anti-realists concerning mathematics. They suggest that, in denying the existence of mathematical objects, anti-realists lose access to a great deal of explanatory reasoning that proceeds in terms of mathematical objects. Since 'explanatory power' is a well-established theoretical virtue, anti-realists who deny the existence of the *mathematical* objects posited by their scientific theories are, according to Baker and Colyvan, left with far inferior accounts of the physical world.

Although I accept that many of our best scientific explanations are mathematical, I wish to argue that the use of mathematics in these explanations is not ontologically committing. This is because of the peculiar role that mathematical theories have in scientific explanation. Since, I will argue, mathematical theories are introduced in empirical science in order to provide models which allow us to *represent* physical systems as having particular physical properties, the question of the existence of the *mathematical* objects posited by these models makes no difference to the utility of our mathematical theorizing. Although there are cases where we might not be able to represent a particular physical phenomena except by relating it to a mathematical model, the utility of the mathematical model in such an explanation does not depend on the actual existence of the objects posited by the model. An understanding of the model as a *theoretical fiction* will account for its role in theorizing equally well, and as such, no explanatory power is lost if we suppose that the mathematical systems made use of in the course of providing mathematical explanations of physical phenomena are mere fictions.

Before I present Baker and Colyvan's argument, and my response, in more detail, I would briefly like to draw attention to some other issues that arise concerning the possibility of explanatory mathematical reasoning. I will not have the space to discuss these issues in any depth, but only mention them so as to highlight the variety of questions that may be raised even before we consider the issue of the ontological commitments incurred in mathematical explanations of physical phenomena.

1 Some Philosophical Issues Surrounding Mathematical Explanation

1.1 The Descriptive Project

Our ontological question presupposes that there are some mathematical explanations of physical phenomena. It is also usually supposed that there are mathematical explanations of mathematical phenomena. (In fact, Mark Steiner has suggested that the existence of the former kind of explanation requires the existence of the latter, since, in cases of mathematical explanations of physical phenomena, "when we remove the physics, we remain with a mathematical explanation—of a mathematical truth!" [Steiner, 1978b, p. 19] Nevertheless, questions have been raised concerning whether it is proper to talk of *mathematical* explanation at all. In the empirical sciences, it is sometimes supposed that all genuine explanations must ultimately be *causal*, so that explaining why event E happened simply involves providing E's causal history. But then, as mathematical objects are usually assumed to be acausal, this would preclude the possibility of any genuine mathematical explanations, at least of physical phenomena. However, as examples such as those provided by Baker and Colyvan show, such an account of explanation can easily be found wanting in some contexts, in particular where explanations are required of general structural facts, rather than of particular instances of particular events. I will discuss these examples of empirical phenomena requiring mathematical explanations later, in part 2.

How about mathematical explanations of mathematical phenomena? Mathematicians often talk of particular proofs, for example, as explaining *why* a particular result is true, and contrast these with those that simply demonstrate *that* the result holds. But is this contrast a genuine one? In the case of explanations of contingent facts, when we ask why an event E occurred, it is often because we realize that, had things been different, E might not have happened, so that, in asking for an explanation for E we are asking for other contingent facts that, together with laws, made it the case that E, rather than, say, F, had to happen. For this reason, it is not sufficient to explain E simply to derive it from some background assumptions. We also need to understand how those assumptions contributed to E occurring, such that, if one or more of them were different, we would have experienced a different outcome.

But the fact that a particular mathematical proposition P follows from a mathematical theory T is a *necessary* truth. As such, we cannot similarly consider what would have had to be different to stop P from 'happening'. Room, then, for a contrast between explanation and mere demonstration seems to disappear: P is true in theory T simply because P follows from the axioms of T, and any proof that shows that this is the case will be equally explanatory. It is, then, arguably a mistake to talk of only some mathematical proofs as being explanatory—perhaps we should say that all are, or none at all.

This line of argument has been countered by Mark Steiner [1978a]. In order to account for the perceived contrast between explanatory and merely demonstrative proofs (proofs that explain versus those that merely convince), Steiner has presented an account of mathematical explanation in terms of which explanatory proofs are those that, by picking out a characterizing property of the class of objects/structures under study, show why a theorem holds for that particular class of objects/structures, and not for those in a related contrast class that do not have that characterizing property. Steiner's approach shows that there is, after all, room for an analogy with our explanations of contingent events. In the empirical case, we're asking what's special about the contingent facts that meant that E, rather than some other event, had to happen, while in the mathematical case, we're asking what's special about the mathematical assumptions we're making that mean that P, rather than a structurally related conjecture Q, can be proved.

Nevertheless, Steiner's approach is not without its critics. Michael Resnik and David Kushner, for example, take issue with Steiner's account on the grounds that Steiner's criterion for an explanatory proof appears neither necessary nor sufficient for mathematical explanation [Resnik and Kushner, 1987]. More interesting, given our worries about the contrast between proofs that convince and proofs that merely explain, Resnik and Kushner argue (with help from van Fraassen's account of explanations as answers to why-questions) that the line between these two classes is highly context dependent. If we simply wish to know why a particular mathematical result holds rather than not, then any proof will count as explanatory for us. If we wish to know why a result holds of structures in class A rather than contrast class B, our explanatory proofs might look more like Steiner's paradigm cases.

More recently, David Sandborg [1998] has argued that even the theory of why-questions relied on by Resnik and Kushner is insufficient to account for all examples of explanatory reasoning in mathematical proofs, since often the best explanations are those that tell us which why-question we should have been asking. Related to this is the more general question of whether any of the standard accounts of explanation that have been proposed within the context of the philosophy of science can account for distinctively mathematical explanations. If there are genuine mathematical explanations, then these surely present a test case for the various accounts of explanation in the empirical sciences, which are often formulated without a view to such forms of explanation.[1]

Clearly, then, there remains much to be said on the question of what, if anything, is to count as an explanatory proof in mathematics. And of course, this doesn't even begin to touch the further questions that arise when we consider the possibility of mathematical explanations that do not take the form of proofs. Let us suppose, though, that there are some mathematical explanations, and that, even if we have no complete analysis of what it is for a piece of mathematical reasoning to be explanatory, we can nevertheless recognize some paradigm examples of mathematical explanations when we see them. This is plausible, even if Resnik and Kushner are right that 'explanatoriness' is a context-dependent, rather than absolute, property of a mathematical proof. We can still accept that, in given

---

[1] Philip Kitcher's work on explanation as unification is a notable exception to this rule, as it is formulated with mathematical as well as physical examples in mind.

contexts, mathematical reasoning that can properly be called explanatory will exist. Supposing that there are some mathematical explanations, what *metaphysical* issues arise out of the existence of mathematical explanations of mathematical and empirical facts?

1.2 The Metaphysical Project

On the question of realism in mathematics, Dummett has cited, approvingly, the claim (which has become known as Kreisel's dictum), that "the problem is not the existence of mathematical objects, but the objectivity of mathematical statements" [Dummett, 1978, p. xxviii, see also p. 146]. I prefer to consider there to be two separate and equally important metaphysical issues here that arise when one considers mathematics. That objectively correct inferences can be made on the basis of explicitly fictional assumptions in literature suggests that objectivity can be achieved without objects. Similarly, in matters of taste, the perceived lack of a 'fact of the matter' concerning, for example, the tastiness of tomatoes suggests that statements about genuine objects can lack objectivity. Hence we should expect that questions concerning the existence of mathematical objects, and those concerning the objectivity of mathematical statements, can be asked independently of one another. Attempts to deal with either one of these issues need not preclude investigation into the other.

In the case of mathematics, though, Dummett has used Frege's context principle to dismiss suspicion with respect to the existence of abstract mathematical objects. Dummett uses Frege's principle, according to which we should "never ... ask for the meaning of a word in isolation, but only in the context of a proposition" (*Grundlagen*, p. x), to draw the conclusion that, "If a word functions as a proper name, then it *is* a proper name." [Dummett, 1956, p. 40]. However, as Dummett himself elsewhere concedes [Dummett, 1978, p. xlii], the context principle understood as a claim about *sense* cannot be applied so easily to draw conclusions about reference. While I accept that the sense of the apparently referring expressions of our mathematical theories is to be given contextually, I see no reason why this should stop us from questioning whether these expressions actually do refer. I thus reserve the right to consider, not only the question of the objectivity of mathematical statements, but also that of the existence of mathematical objects.

In fact, the existence of mathematical explanations has been used in different ways to support realism in both of these senses. Although I am most interested in this paper in considering the inference from the existence of mathematical explanations to the existence of mathematical objects, I would also like to say a few words concerning the use of mathematical explanations to support claims concerning the objectivity of mathematics.

Realism (1): Objectivity

Does the existence of mathematical explanations of mathematical phenomena speak to the objectivity of mathematics? Oddly enough, the very same example of such an explanation has been used quite independently by Freidrich Waismann [1982] and by Mark Steiner [2000] to answer this question in the affirmative.

Reacting against a Wittgensteinian line of thought according to which new proofs serve to *create* new mathematical connections (so that, "even if the proved mathematical proposition seems to point to a reality outside itself, still it is only the expression of acceptance of a new measure of reality" [Wittgenstein, 1978, §27]), Waismann asks us to consider the following Taylor series expansion, for $x \in \mathbb{R}$:

$$\frac{1}{1+x^2} = 1 - x^2 + x^4 - x^6 + \ldots$$

This expansion converges for $|x| < 1$, but diverges for all other real values of x. Why is this?

If we stick to the real numbers, no answer presents itself. To find a mathematical explanation of the behaviour of this function, we need to consider it as a function over the complex numbers. Consider, then, the function

$$\frac{1}{1+z^2},$$

for $z \in \mathbb{C}$. Clearly, this function has singularities at $z = \pm i$, for here the denominator is zero. But this means that no Taylor series expansion could converge at these values of z. Furthermore, it is a theorem of complex analysis that any power series expansion only converges within a circle of radius R, for some R, about the origin, and diverges elsewhere. Hence, for $|z| \geq 1$, the expansion cannot converge. The singularity at $z = \pm i$ thus precludes even the expansion of the real-valued function from converging for $|x| \geq 1$. It is, as Waismann remarked, as if the real function already knew that the complex numbers were there.

For Waismann, this use of the complex numbers to explain a known fact about the reals is used to support the non-arbitrariness of our extension of the number system to complex numbers. Furthermore, the objectivity of our inferences concerning the real numbers receives support here when we discover a new way of accounting for a previously inferred statement. Another angle on objectivity is presented by Steiner, who makes a slightly different use of the same example to infer the objectivity of our representations of complex numbers. According to Steiner, the statement about the convergence of the expansion to the real valued function is in need of an explanation. But in order to provide the tools required for such an explanation, we need to set up the system of complex numbers, together with the geometric representation of complex numbers, in order to provide us with the notion of a radius of convergence. The fact that such concepts are *required* to explain this previously known fact gives a air of objectivity to our decision to develop the complex numbers and their geometric representation.

Since my main interest in this paper is in the case for realism about mathematical objects, and not the case for the objectivity of mathematical statements, there is no space here for a critical examination of such arguments from explanatoriness to objectivity. I am inclined to agree with Steiner and Waismann that examples such as theirs, where an 'old fact' is explained using the machinery of a new mathematical system do provide a striking case for some kind of objectivity in our mathematical theories. But let us leave these examples and consider a further use of mathematical explanations, to provide evidence for the existence of mathematical objects.

Realism (2): Objects

Recently, Mark Colyvan [2002; ms], and Alan Baker [2005], have both used the existence of mathematical explanations of *physical* phenomena to support realism about mathematical objects. In a nutshell, the argument is as follows:

- (P1) Genuine explanations must have *true* explanans. In particular, then, the objects posited by those explanans must exist.

- (P2) There are some genuine mathematical explanations of some physical phenomena.

Therefore

- (C) The mathematical objects posited by these explanations must exist.

This argument presents a new twist on the Quine-Putnam *indispensability argument* for the existence of mathematical objects, which they claim avoids problems that have been found with that argument. While I accept, with Baker and Colyvan, that there are some genuine mathematical explanations of physical phenomena, I want in the second part of this paper to deny that this has the ontological consequences that they suppose. I will therefore be questioning P1, the claim that all good explanations must have true explanans.[2]

2 Mathematical Explanations and the Reality of Mathematical Objects

2.1 Some Background

The Original Indispensability Argument

It is often objected that philosophers who wish to engage in ontology are guilty of "introducing a philosophical sense of 'exists' which is distinct from the ordinary application of 'there is...'" [Dummett, 1956, p. 40]. Rudolf Carnap's attitude to conflicts over existence questions in this philosophical sense was to regard them as ill-conceived, at least "until both parties to the controversy offer a common interpretation of the question as a cognitive question; this would involve an indication of possible evidence regarded as relevant by both sides." [Carnap, 1956, pp. 218–9] In raising questions about the existence of mathematical objects, it is therefore important to make clear to what sense of existence one is appealing.

Here I take my cue from W. V. Quine, who sought to reclaim the philosophical project of ontology by insisting on a sense of 'exists' that takes its cue from our ordinary use of the word. For Quine, starting from our common sense understanding of concepts such as existence, there is a continuum which leads to empirical science, the result of our best efforts at applying

---

[2] Given the form of Baker and Colyvan's argument, one might wonder why it is mathematical explanations of *physical* phenomena that get priority. For if there are, as we have suggested, some genuine mathematical explanations of mathematical phenomena, then these explanations must also have true explanans. The reason that this argument can't be used is that, in the context of an argument for realism about mathematics, it is question-begging. For we also assume here that genuine explanations must have a true *explanandum*, and when the explanandum is mathematical, its truth will also be in question.

and refining these concepts, and to philosophy as the continued application of scientific methods to further questions that arise. Thus, according to Quine,

> science is self-conscious common sense. And philosophy in turn, as an effort to get clearer on things, is not to be distinguished in essential points of purpose and method from good and bad science. [Quine, 1960, pp. 3–4]

Our philosophical inquiry into questions of existence is legitimate to the extent that it takes its cue from good science. In particular, for Quine, we discover our ontological commitments by looking to the commitments of our scientific theories. We can label this approach to ontology 'ontological naturalism'. Its central thesis is that we should look to science to tell us what there is.

According to Quine, ontological naturalism leads quickly to realism concerning mathematical objects. This is because our best scientific theories indispensably posit the existence of mathematical objects, so trusting those theories involves accepting the existence of the mathematical objects they posit. Central to this argument is a hidden premise of confirmational holism, according to which we cannot pick and choose which parts of our scientific theories to believe, but must instead accept that, if our theories receive confirmation at all, that confirmation extends to all of their posits. The argument (known as the indispensability argument) can thus be presented as follows:

1. Positing mathematical objects is indispensable in formulating our best scientific theories. (*Indispensability*)

2. We should look to science to tell us what there is. (*Ontological Naturalism*)

3. The confirmation our scientific theories receive extends to all of their posits, mathematical and physical, equally. (*Confirmational Holism*)

   Therefore:

4. We ought to believe in the existence of those mathematical objects posited by our scientific theories.

According to this argument, then, the mere mention of an object in the course of our scientific theorizing is evidence for existence. By confirmational holism, confirmation extends to all theoretical posits, whether or not they appear in *explanations*, so the issue of mathematical explanation is not particularly relevant to this argument for realism. However, confirmational holism has recently been questioned by theorists who, nevertheless, accept the other premises of the argument. By bringing in the indispensable role of mathematical posits in scientific *explanations*, Baker and Colyvan hope to show that, even if confirmational holism is incorrect, ordinary scientific standards support the claim that mathematical objects receive confirmation through their role in our scientific theories. We thus have a new wrinkle on the indispensability argument.

Troubles with Holism

Objections to confirmational holism, purporting to undermine the original indispensability argument, have been presented by Penelope Maddy [1992; 1997]; Elliott Sober [1993]; Joseph Melia [2000]; and Mark Balaguer [1998], amongst others. Both Maddy and Sober suggest that attention to ordinary scientific practices shows that not all components of our scientific theories are considered to be under test when these theories are tested, and not all those components should be considered confirmed when our theories receive confirmation. Maddy points to cases where our theories make use of explicit idealizations, not all of which can be replaced by literally true alternative theories. The success of such a theory does not license our belief in the truth of the idealized components. Furthermore, even in theories that are not intended as explicit idealizations, it takes more than a mere mention in our scientific theories to convince scientists of the reality of a theoretical posit. Maddy gives examples of scientists reserving judgment on the question of the existence of some of the objects posited by their theories, until they have some kind of 'direct evidence' of existence.

Sober also sees a distinction between those theoretical posits that receive confirmation in testing and those that do not. Stressing the comparative aspect of theory testing, whereby hypotheses are tested against a backdrop of competitors, Sober argues that our standard empirical tests provide no test of the pure mathematical hypotheses utilized by our theories, since those hypotheses form a backdrop for all the competing theories under test. According to Sober, "If the mathematical statements M are part of each hypothesis under test, then the observational outcome does not favor M *over any of its competitors.*" [Sober, 1993, p. 45] It might be objected that we can set up experiments to test particular mathematical hypotheses against competitors, for example by setting up arrays of physical objects to test rival hypotheses about addition and multiplication. But Sober points out here that, if we do count two apples added to two more apples and find ourselves with three, rather than four, as an answer, we'd always look for some problem with the 'experiment', rather than change our mind about the sum in question. But if we hold the belief that $2 + 2 = 4$ immune from revision, we are not putting it to genuine test in this experiment, and cannot, therefore, claim empirical support in this case.

At any rate, the hypothesis about mathematical objects we want to test is the hypothesis that such objects exist. The relevant contrasting hypothesis is then that there are no such objects, and that, therefore, our empirical theories contain fictional components and are at most correct in their non-mathematical claims. But Mark Balaguer has suggested empirical evidence can never decide between these two global hypotheses concerning mathematical objects, since, given that mathematical objects, if they exist, are causally isolated from empirical objects, the two hypotheses (that our empirical theories include true descriptions of the mathematical realm and that our theories include mathematical components that are merely fictional) do not differ in any of their empirical implications.[3]

We have, then, a suggestion first of all that not all theoretical components are equally under test when we test our scientific theories, and second, that in particular the mathematical components of our theories are not put to test. Furthermore, if Balaguer is right, then the mathematical existence assumptions made in our theories *could* never be meaningfully tested. Joseph Melia's contribution to this picture is to present an account of the role of mathematical hypotheses in our theories that suggests an explanation of *why* they are apparently essential to empirical science, even though empirical science might not require these hypotheses to be true. Melia argues that mathematical language will sometimes be indispensable in providing us with a way of presenting configurations of recognisably concrete systems. It is reasonable, then, to think of the mathematical hypotheses present in our theories as convenient fictions which give us the means to represent physical

---

[3]Balaguer puts this point in a rather more snappy, but slightly misleading, way as follows: "We can think of it this way: if all the objects in the mathematical realm suddenly *disappeared*, nothing would change in the physical world; thus, if empirical science is true right now, then its nominalistic content would remain true, even if the mathematical realm disappeared; but this suggests that if there never existed any mathematical objects to begin with, the nominalistic content of empirical science could nonetheless be true." [Balaguer, 1998, p. 132]

systems of physical objects. That these hypotheses are present merely for representational utility, and not as working hypotheses about mathematical reality, is suggested in the practice amongst scientists of formulating their theories mathematically but denying the existence of mathematical objects:

> Philosophers typically represent these scientists as engaging in double-think—denying by night what they believe by day. But it is surely uncharitable to regard so many scientists as hypocrites! Surely it is more charitable to think that we must have misinterpreted them.
>
> But look at the kind of things they say: "The force between two massive objects is *proportional* to the *product* of the masses *divided* by the square of the distance"; "There is a one-to-one differentiable *function* from the points of space-time onto *quadruples of real numbers*"—how can we have misinterpreted them? By thinking that any theorist who presents a theory of the world must do so by asserting a set of sentences, each one believed by the theorist. This is our mistake. As soon as we allow theorists to take away details that were added before, to subtract parts of their earlier discourse, the theorists no longer appear to believe contradictory things. The mathematics is the necessary scaffolding upon which the bridge must be built. But once the bridge has been built, the scaffolding can be removed. [Melia, 2000, p. 469]

We can draw from these discussions the following response to the original indispensability argument:

- Not all scientific posits are equal.

- In particular, the role played by mathematical posits in our theories is quite different from the role played by physical posits. (Mathematical objects are included in our theories to provide us with means of representing how things are with physical systems.)

- This difference in role makes it reasonable to take a different attitude to mathematical posits than we do to physical posits. (In particular, we can account for the representational role of mathematical posits on the hypothesis that they are mere fictions, while we cannot account for the causal role of electrons, for example, on this hypothesis.)

- We ought to accept, but not believe, our scientific theories. We accept them as the best available representations of physical phenomena, but we do not think that we have reason to believe that all the objects we posit in the process of providing such representations exist.

Against the Quinean view, then, the central claim of this response is that we are not committed to belief in the existence of objects posited by our scientific theories *if their role in those theories is merely to represent configurations of physical objects.* Fictional objects can represent just as well as real objects can.

A Realist Response

I will not say any more in defence of this position, but will rather consider how Baker and Colyvan's shift to the role of mathematical posits in explanations attempts to respond to this view. It is important to note that both Baker and Colyvan accept, for the sake of argument at least, that these considerations work to undermine the indispensability argument in its original form. That is, they are prepared to reject confirmational holism and to accept for now that the confirmation of our scientific theories would not extend to those components whose presence in our theories is merely to represent physical configurations of physical objects.

Baker and Colyvan's strategy is then to argue that mathematical posits are not present in our theories merely in order to represent, or model, physical phenomena. Rather, they argue, mathematical posits have much the same role as physical posits. In particular, mathematical objects are often introduced for their *explanatory power*. Although supposing that mathematical objects do not exist makes no difference to the predictions our theory makes concerning empirical matters, this does not, according to Baker and Colyvan, mean that there is no reason to prefer the hypothesis that mathematical objects exist to the hypothesis that they do not. We should prefer the hypothesis that mathematical objects exist because, without this hypothesis, our theories suffer a loss in *explanatory power*. So ordinary scientific grounds for theory choice speak in favour of the hypothesis of the existence of mathematical objects.

The success of such a response depends on two components. Firstly, Baker and Colyvan must show that there are some good scientific explanations that posit mathematical objects, and secondly, they must show that these explanations cease to be good explanations on the hypothesis that there are no such objects. In fact, both Baker and Colyvan focus on establishing only the first of these two claims, and simply assume that all good explanations must have true explanans, so that if we drop the assumption that the mathematical objects posited by our explanations exist, then these 'explanations' cease to explain at all.

Now, it is widely accepted that good explanations must have true explanans (but see [van Fraassen, 1980] and [Cartwright, 1983] for voices of dissent), and so it is not surprising that Baker and Colyvan see themselves as free to make this assumption. However, given that Baker and Colyvan have (at least tentatively) accepted that our theories as a whole need not be true to be good, that they may make use of some false hypotheses in order to represent truths about physical systems, it is reasonable to ask whether such an attitude warrants a change in view also of those parts of our theories that function as explanations. If one is willing to accept that literally false theories can still serve to provide accurate representations of physical systems, and that these theories get their value because of the conditions they impose on the behaviour of these systems,[4] then it is plausible that one should hold a similar attitude to theoretical explanations. Couldn't a mathematical explanation get its value *as an explanation* due to the conditions it imposes on concrete, non-mathematical systems? And couldn't these conditions be imposed equally well by a fictional theory as they would be by a literally true one?

I would like to argue that the examples of mathematical explanations given by Baker and Colyvan, like scientific theories in general, can be understood as valuable (and in particular as *explanatory*), even on the assumption that the mathematical objects posited by those explanations do not exist. To see this, let us consider three examples of slightly different cases of mathematical explanations of physical phenomena that Baker and Colyvan have presented.

---

[4]How can a false theory impose any conditions on the behaviour of a physical system? Well, if true, such a theory would be true in virtue of mathematical objects being configured in a certain way and physical systems being configured in a certain way, so as to allow for the various relations posited between the mathematical and physical components to hold. The condition imposed on the physical world by such a theory is, as Mark Balaguer puts it, "that the physical world holds up its end of the "empirical-science bargain"" [Balaguer, 1998, p. 136].

2.2 Mathematical Explanations of Physical Phenomena: Some Examples

Antipodal Weather Patterns [Colyvan, 2001]

Mark Colyvan [2001] presents the following example: Suppose that two antipodal points, p_1 and p_2, on the surface of the earth are found to have exactly the same barometric pressure and temperature. Why is this? One explanation could be given in purely causal terms, citing the causal history of each point, and thus explaining how it is they got to be the pressure and temperature they are now at. But something is left out of this explanation. We are told why p_1 has the pressure and temperature it has, and why p_2 has the pressure and temperature *it* has, but not why both of these opposite points have the same pressure and temperature as each other. The causal history thus appears wanting.

To find an explanation of why these two points have the same pressure and temperature as each other, we need to look to mathematics, and in particular to the *Borsuk-Ulam theorem* of algebraic topology. According to this theorem, every continuous function $f : \mathbb{S}^n \to \mathbb{R}^n$ maps some pair of antipodal points to the same point. Noting that the earth is topologically equivalent to the sphere $\mathbb{S}^2$, and treating temperature and pressure as continuous functions on the earth's surface, it follows that there will always be antipodal points p_1 and p_2 at the same pressure and temperature. It appears, then, that the mathematical theorem provides the explanation we wanted of this fact.

A couple of points are worth noting here. First of all, as Alan Baker has pointed out [Baker, 2005], Colyvan's example is somewhat problematic in that it is highly manufactured. Colyvan envisages the discovery of p_1 and p_2 sparking a search for an explanation, which is then satisfied by the Borsuk-Ulam theorem. But in fact, no such discovery has been made (and it is extremely unlikely that we'd ever be able to, let alone wish to, catalogue pressure and temperature values to the extent and level of detail required to come up with two such points). Colyvan's imagined example is driven by the mathematical result, which, together with the linking assumptions already mentioned, makes a prediction about pressure and temperature measures on the earth's surface. Secondly, the assumptions required for such a result to be compelling, including the assumption that pressure and temperature can be treated as continuous functions taking individual space-time points to real numbers, are themselves idealizations that likely break down when probed too closely.[5] If this is right, then the explanans of this explanation are already not all literally true, even if we assume the existence of real numbers and functions.

But let us set aside these worries and imagine for now that this is a genuine explanation, in the sense that, had we in fact discovered p_1 and p_2 as in the example, we *would* accept this account as an explanation of their pressure and temperature properties. The main purpose of this example for Colyvan's argument is to show that there are cases where causal explanations are not enough, as against those who would insist that the only way we can explain a fact is to trace its causal history. I am inclined to agree with Colyvan that this example presents a *prima facie* counterexample to such a view. But the question we need to consider is, what makes this explanation explanatory? In particular, is the hypothesis of the existence of mathematical objects essential for us to be able to see this as a genuine explanation of the phenomenon in question?

What makes this explanation work? We model the earth as a sphere, and pressure and temperature as continuous functions on the surface of this sphere. Once we have done this, the Borsuk-Ulam theorem can be seen to apply, and, to the extent that are model is a good one, we can draw a conclusion about the existence of a pair of points on the earth's surface. Does the question of whether the sphere and the functions in our model *really exist* matter to the success of this piece of reasoning? It is hard to see how it should. What matters is that the mathematics used provides a good model of relevant features of the earth. But, as Colyvan and Baker have already conceded, a literally false theory may be able to provide such a model as well just as a true one. If one concedes that literally false mathematics may nevertheless correctly represent features of physical systems, then the role of the mathematical hypotheses in this *explanation* present no new problems for the anti-realist view. The explanation works because it provides a correct representation of the physical system in question, not because all of its hypotheses are literally true.

Circle-squaring [Colyvan, ms]

A second example of Colyvan's, in [Colyvan, ms], appears initially to avoid this problem, in that it appears to appeal directly to mathematical properties of mathematical objects to explain an empirical phenomenon. Colyvan's

---

[5] I am grateful to Brian Davies for drawing my attention to this point.

example is circle-squaring: it is an empirical fact about our constructing abilities that, using only a ruler and compass, we cannot manage to construct a square of the same area as a given circle. Why is circle-squaring impossible? Because π is transcendental. The explanation of the empirical phenomenon in this case appeals to a property of a mathematical object. Surely the assumption of the existence of the object in question is necessary for our ability to provide such an explanation?

Again, Colyvan's example is somewhat idealized: we do not have empirical evidence that we can *never* construct a square of the same area as a given circle, but at best only that we have not done so far.[6] What we do have is a mathematical theorem that *predicts* that we will never find such a construction, and which alternatively can be used to explain why we haven't managed such a construction so far (even in those cases where it may look to us like we have managed such a construction). But let's set this quibble aside, and assume again that we have an explanation to the extent that, if we *had* realized that our construction attempts never led us to square the circle, appeal to the transcendentality of π would explain this.

But although the short form of the explanation is, as Colyvan says, 'because π is transcendental', the fact that this works as an explanation depends again on setting up a mathematical model. Physical drawings of circles and squares, and the physical constructions possible using the instruments of a ruler and a pair of compasses, are represented in geometry as their idealized counterparts: perfect circles and squares; an infinitely long and perfectly straight edge, and an infallible devices for drawing a circle of any radius about a chosen point. Then, it is shown that the constructions that can be made using these instruments are equivalent to algebraic operations of addition, subtraction, multiplication, division, and taking square roots. But for a given circle, of radius r and area πr^2, the construction we need, given r, is of a square whose edge is $\sqrt{\pi}r$. Of course, given π, we could construct $\sqrt{\pi}$ and therefore $\sqrt{\pi}r$. But, since π is a transcendental

---

[6]I say 'at best' here because it is plausible that, without knowledge of the mathematical proof of the impossibility of circle-squaring, we would not know that none of our attempts at circle-squaring haven't succeeded. As mathematicians will know, just as crank inventors occasionally announce that they have succeeded in inventing perpetual motion machines, so crank mathematicians are fond of announcing circle-squaring constructions. For them, the empirical evidence is that they have managed to square a circle. For mathematicians presented with these examples, it is the mathematical impossibility result that overturns this evidence.

number, it cannot be constructed using only the algebraic operations mentioned above, and so we cannot construct the square edge required. This is the sense in which the transcendentality of π explains the impossibility of the construction in question.

Again, then, we have an idealized mathematical model that represents the physical processes in question. To the extent that the model is a good model of these processes, then the mathematical impossibility result can be converted to an impossibility result regarding the physical construction of a square the area of a given circle using the tools provided. Certainly, in the mathematical explanation of the mathematical result, it is because our mathematical theory tells us that π is transcendental that we know that we cannot make the *mathematical* construction in question. But we have already (see footnote 2) dismissed as question-begging attempts to take mathematical explanations of mathematical results as having ontological implications. Furthermore, no ontological mileage can be got from the use of the mathematical impossibility result to tell us about a physical impossibility, at least not by theorists such as Baker and Colyvan, who accept that mathematical systems can correctly represent aspects of physical phenomena even if the objects posited by those systems do not in fact exist. As with the previous example, the effectiveness of this explanation can be put down to the correctness of the model as a representation of the physical system, and not to the truth of the mathematics involved in this model.

Periodical Magicicada Cicadas [Baker, 2005]

The final example I want to consider is Alan Baker's [2005]. Baker presents this example in part because, unlike the examples of Colyvan's so far considered, it is a real-life case of a felt need for scientific explanation being answered by a mathematical result. Baker's example is of the magnificently named Periodical Magicicada Cicadas. According to Baker, these North American insects remain at the nymphal stage, in the soil, for thirteen or seventeen years (depending on the geographical region).[7] The adults then all emerge within the same few days of the same year, mate, and then die, their offspring restarting the cycle. One question that can be asked is, why do they have the period (13 or 17) that they have?

Some biological assumptions are necessary here. Firstly, a relatively long

---

[7]The seventeen-year variety have recently been in the news since they emerged, as expected, in the summer of 2004.

life-cycle is required in order to allow the nymphs to reach maturity, due to the low levels of nutrients in the soil and the low soil temperatures for much of the year. The shorter life-cycle insects are those found in southern parts of the U.S., where soil temperatures do not drop so low and so nymph development can happen more quickly. Secondly, since it is important that, after development, the insects find a mate, a periodical life cycle makes sense, since there is most chance of finding a mate if all the nymphs are synchronized. But why are the periods the lengths they are (13 or 17 years) in particular? Here, Baker tells us, biologists have two competing hypotheses. First, they postulate the existence, in the evolutionary history of the cicadas, of periodically appearing predators. Clearly, it would be advantageous for the cicadas to time their appearance so as to avoid appearing in years where these predators appear. Alternatively, it is suggested that there may also have been other periodical creatures similar enough to the Magicicada Cicadas, to be able to mate with the Cicadas. In this case, again it would be advantageous for the cicadas to time their appearance to avoid these similar creatures: if not, they might mate without producing viable offspring, or alternatively, the offspring they did produce would have a period somewhere between the lengths of the two periods of the parents. In this case, they would appear in a year where, potentially, no other mates would appear. So, on each of these hypotheses, it becomes advantageous to the Cicadas to have a cycle length that will overlap minimally with other periodical creatures.

With this background in place, the explanation of the length of period becomes a simple one: the Magicicada Cicadas have settled on the periods they in fact have *because* 13 *and* 17 *are prime numbers*. To see this, suppose that two periodical species, S_1 and S_2, of period lengths m and n respectively, overlap in some year, let's say year 0. Then the next time they will overlap will be at year l, where l is the lowest common multiple of m and n (l.c.m(m,n)), and they will then intersect every l years. At best, l.c.m(m, n) will simply be $m \times n$. In fact, it is easy to show that this happens only when m and n have no common factors aside from 1—that is, when they are coprime. So to avoid each other as much as possible, S_1 and S_2 should 'pick' periods that are coprime. Now, for species S_1 to maximize its chances of avoiding many other periodical species, it should 'pick' a period that is coprime to the most numbers close to it. But a prime number m is coprime to all numbers less than $2m$, so it is advantageous for S_1 to choose a prime period.

This is an impressive example of a genuinely mathematical explanation of a previously noticed empirical phenomenon. Furthermore, it proceeds in terms of properties of natural numbers: it is the primeness of 13 and 17 which makes them good choices of period for the cicadas. Indeed, it could be said that the cicadas have discovered for themselves a bit of number theory, and are exploiting a bit of mathematical knowledge to their advantage. Surely, if any example is going to make Baker and Colyvan's point, this one will?

But again, we need to ask if anything is lost in the explanation of cicada behaviour if we drop the assumption that numbers exist. And once more, in this case it is plausible that nothing is lost. Why is it that the primeness of 13 can serve as an explanation of the cicada behaviour? Only because the succession of years is correctly modelled by the natural number system. Starting from the first year at which the cicadas overlap with some other cyclic creature, and labelling the successive years 1st, 2nd, 3rd..., we find we have an ω-sequence (or at least, the initial segment of a potential ω-sequence). But now, we can account for the applicability of the mathematical result not as due to there actually being a number 13 or 17 with the property of primeness, but rather because, it follows from the assumptions of number theory, that 13 and 17 are prime. To the extent that the sequence of natural numbers provides a good model of the sequence of years since some first, 'overlapping' year, that is, to the extent that this sequence of years can be considered as structurally isomorphic to an ω-sequence, then we should expect that facts about what follows from the assumptions of number theory will be relevant to facts about relations between these years. Nothing here requires there actually to be a sequence consisting of 'the' natural numbers, or even that there is any completed ω-sequence. So nothing is lost in the explanation of cicada behaviour if we drop the assumption that natural numbers exist.

3 Conclusion

Baker and Colyvan present their argument, from the existence of mathematical explanations of physical phenomena to the existence of mathematical objects, to support a new version of the 'indispensability' argument. The original Quine-Putnam indispensability argument is, they concede, vulnerable to the objection that it is a distortion of our scientific theories to consider all components of those theories to be equally confirmed by theoretical successes. Some theoretical assumptions are introduced in order to allow us the linguistic resources to represent certain aspects of physical systems, and it would be a mistake to interpret those aspects of our theories as literally true, when there is no evidence that scientific practice warrants such an interpretation. Baker and Colyvan therefore concede that scientific theories might provide valuable representations of the world without being literally true in all of their component parts.

Having conceded this, though, Baker and Colyvan return to a more traditional realist position in assuming that, though scientific theories as a whole need not be true to be good, scientific *explanations* ought to be. Hence, their new 'indispensability' argument depends on finding indispensable use of mathematical assumptions in scientific explanations. But if the original indispensability argument can be rejected on the grounds that some theoretical components can be good representations without being true (so that 'fictional' assumptions would do the representative work just as well), then the same considerations can be applied in the case of theoretical explanations. In fact, in the cases they have presented, no explanatory value appears to be lost if we assume only (a) that the mathematical assumptions in the explanations presented serve to provide a good model of the physical phenomenon in question; and (b) that the mathematical reasoning used in these explanations is objective (so that we can expect that conclusions drawn about the models will also apply to the systems modelled). But the objectivity of mathematical reasoning is not in question here—in fact, we have suggested that the existence of mathematical explanations of *mathematical* phenomena supports the various hypotheses concerning the objectivity of mathematics. What is not supported by the examples of explanatory mathematical reasoning presented in this paper is the hypothesis of the existence of mathematical objects.

Acknowledgements

Aside from the generous and perceptive comments of audience members at the Rome conference, I am also grateful to Alan Baker and Mark Colyvan for their comments on a draft of this paper.

BIBLIOGRAPHY

[Baker, 2005] Alan Baker. Are there genuine mathematical explanations of physical phenomena? *Mind*, 2005. (forthcoming).

[Balaguer, 1998] Mark Balaguer. *Platonism and Anti-Platonism in Mathematics*. Oxford University Press, Oxford, 1998.

[Carnap, 1950] Rudolf Carnap. Empiricism, semantics and ontology. *Revue Internationale de Philosophie*, 4:20–40, 1950. Revised and reprinted in [Carnap, 1956, pp. 205–221].

[Carnap, 1956] Rudolf Carnap. *Meaning and Necessity: A Study in Semantics and Modal Logic*, 2nd Edition. University of Chicago Press, Chicago, IL, 1956.

[Cartwright, 1983] Nancy Cartwright. *How the Laws of Physics Lie*. Oxford University Press, Oxford, 1983.

[Colyvan, 2001] Mark Colyvan. *The Indispensability of Mathematics*. Oxford University Press, Oxford, 2001.

[Colyvan, 2002] Mark Colyvan. Mathematics and aesthetic considerations in science. *Mind*, 111:69–74, 2002.

[Colyvan, ms] Mark Colyvan. Mathematical recreation versus mathematical knowledge. ms. Forthcoming in [Leng, Paseau and Potter, forthcoming].

[Dummett, 1956] Michael Dummett. Nominalism. *Philosophical Review*, 65:491–505, 1956. Reprinted in [Dummett, 1978, pp. 38–49].

[Dummett, 1978] Michael Dummett. *Truth and Other Enigmas*. Harvard University Press, Cambridge, MA, 1978.

[Frege, 1968] Gottlob Frege. *The Foundations of Arithmetic*. Northwestern University Press, Evanston, Ill., 1968. Translation by J. L. Austin of *Die Grundlagen der Arithmetik*, (1884).

[Grosholz and Breger, 2000] Emily Grosholz and Herbert Breger, eds. *The Growth of Mathematical Knowledge*, Kluwer Academic Publishers, Dordrecht, 2000.

[Leng, Paseau and Potter, forthcoming] Mary Leng, Alexander Paseau and Michael Potter, eds. *Mathematical Knowledge*, forthcoming.

[Maddy, 1992] Penelope Maddy. Indispensability and practice. *The Journal of Philosophy*, 89:275–289, 1992.

[Maddy, 1997] Penelope Maddy. *Naturalism in Mathematics*. Clarendon Press, Oxford, 1997.

[Melia, 2000] Joseph Melia. Weaseling away the indispensability argument. *Mind*, 109:458–479, 2000.

[Quine, 1960] Willard Van Orman Quine. *Word and Object*. MIT Press, Cambridge, MA, 1960.

[Resnik and Kushner, 1987] Michael D. Resnik and David Kushner. Explanation, independence and realism in mathematics. *British Journal for the Philosophy of Science*, 38:141–158, 1987.

[Sandborg, 1998] David Sandborg. Mathematical explanation and the theory of why-questions. *British Journal for the Philosophy of Science*, 49:603–624, 1998.

[Sober, 1993] Elliott Sober. Mathematics and indispensability. *Philosophical Review*, 102:35–57, 1993.

[Steiner, 1978a] Mark Steiner. Mathematical explanation. *Philosophical Studies*, 34:135–151, 1978.

[Steiner, 1978b] Mark Steiner. Mathematics, explanation, and scientific knowledge. *Noûs*, 12:17–28, 1978.

[Steiner, 2000] Mark Steiner. Penrose and platonism. 2000. In [Grosholz and Breger, 2000, pp. 133–141].

[van Fraassen, 1980] Bas van Fraassen. *The Scientific Image*. Clarendon Press, Oxford, 1980.

[Waismann, 1982] Freidrich Waismann. *Lectures on the Philosophy of Mathematics*. Rodopi, Amsterdam, 1982.

[Wittgenstein, 1978] Ludwig Wittgenstein. *Remarks on the Foundations of Mathematics*. Blackwell, Oxford, 3rd edition, 1978.

Can a Proof Compel Us?

CESARE COZZO

1 The idea of a compulsion of proof.

The compulsion of proofs is an ancient idea, which plays an important role in Plato's dialogues. The reader perhaps recalls Socrates' question to the slave boy in the *Meno*: "If the side of a square A is 2 feet, and the corresponding area is 4, how long is the side of a square whose area is double, i.e. 8?". The slave answers: "Obviously, Socrates, it will be twice the length" (cf. [Plato, ME, 82–85]). A straightforward analogy: if the area is double, the side is double. Nevertheless, the answer is wrong. Socrates wants to lead the slave to the right conclusion. The boy should reach the truth through steps that are all "his own", performed with full conviction. To this aim, Socrates addresses a series of short pressing questions to the slave boy. Simple questions provoke equally simple replies, though the boy is sometimes puzzled and surprised by the answers he feels compelled to give. In the first part of the exchange the boy is gradually forced to admit that a square B with double side (i.e. side of length 4) has an area which is *not* double, but four times as big, i. e. 16. In the second part the boy hazards the guess that the square with double area has a side of length 3, since 3 is between 2 and 4. But Socrates easily drives him to acknowledging that if the side is 3, the area of the resulting square C is 9, not 8. The boy can only exclaim: "By Zeus, Socrates, I do not know!". In the third part, at Socrates' prompting, the diagonal of the original square A is drawn within the second fourfold square B. Responding to Socrates' questions, the boy is led to conclude that the square D whose side is the diagonal of A is precisely the required square: its area is 8. The exchange between Socrates and the slave boy is a process trough which a symbolic configuration is built up, a mathematical proof. The proof forces the boy to come to a conclusion against his previous conviction. It is instructive that the first answer to which the boy is spontaneously inclined, and which he initially gives, is a wrong answer. To lead him to the right answer, Socrates must elicit from the boy other intermediate unsuspected answers through a skilfully chosen series of questions. In this exchange we clearly see exemplified a kind of power that symbolic constructions have over us. Through a symbolic construction we can change the convictions of our fellow speakers. The compulsion of proofs is a special power proofs have to change beliefs or convictions.

In Plato's *Gorgias*, however, Socrates distinguishes between the compelling force of arithmetic or geometry [Plato, GO, 450–451], which leads to knowledge, and the power of the speeches of an orator "instilling persuasion in the souls of an audience" [Plato, GO, 453] in a law court, in a council meeting or in an assembly. In both cases, mathematics and rhetoric, we attain persuasion, but of two different kinds. The orator employs devices that produce a type of persuasion "providing conviction without knowledge", a persuasion which is reached through *kolakeia*, flattery [Plato, GO, 463]. The orator complies with the moods of the audience and exploits them. His ability is only a knack for producing a certain gratification or pleasure. Socrates compares it to pastry baking. On the other hand, arithmetic or geometry give us a persuasion "providing knowledge" [Plato, GO, 454]. Socrates distinguishes between disputes where the participants are "eager to win instead of investigating the subject under discussion" [Plato, GO, 457], and disputes where participants are interested in truth. In disputes of the latter kind a stronger binding force is attached to arguments which, like geometrical proofs and unlike rhetorical speeches, provide both conviction and knowledge: they are described by Socrates as "arguments of iron and adamant" [Plato, GO, 509]. If we aim at truth, we ought to let us be bound by them. After Plato, the notion of a compelling argument runs through the whole history of philosophy. Arguments of various kinds enjoy the property of being compelling, but mathematical proofs are often considered by philosophers the clearest instance of this property.

Two features of the proof construction in the *Meno* are noteworthy. On the one hand we notice the compulsion by which the boy is driven to some of his answers: he is forced by Socrates' questions to give answers that lead him to an unsuspected conclusion. On the other hand, we notice the very able way in which Socrates chooses the questions. The choice of questions, and the suggestion to draw the diagonal, are moves that, though legitimate, are not forced by the previous steps in the construction of the proof. The example seems to show that we should avoid two opposite mistakes. A first mistake is to focus only on the compulsion of proofs, a mistake which worsens if we interpret compulsion as a property conferring absolute certainty to the whole proof and to its conclusion. In our times the latter view was vividly criticized by Imre Lakatos. He highlighted that actual informal proofs are tentative and fallible, described them as "thought-experiments" [Lakatos, 1978, p. 65] and emphasized that "mathematical heuristic is very like scientific heuristic" [Lakatos, 1976, p. 74]. The exchange between Socrates and the slave shows that an account of mathematical proofs which focuses only on the compulsion of proofs is one-sided, because it neglects the heuristic skill in detecting an efficacious order in which to put questions so as to reach the solution of the given problem. An opposite mistake, however, is to neglect the compulsion of proof. The choice of questions would loose all interest, if the answers given were arbitrary answers. Answers must be specially powerful if their final result has to be a change of our beliefs which clashes with previous convictions and is sometimes not only surprising and unforeseen, but even undesired. Answers perceived as arbitrary would be powerless. But the answers accepted in the end are not arbitrary: they are cogent. We perceive them as leading us to the true conclusion. The slave boy is compelled to reject his initial wrong answer that, if the area is double, the side is double, and is compelled to accept the right solution. About his best known example of thought experiment, Cauchy's proof of Euler's formula on polyhedra, Lakatos writes that, though it is not a concatenation of formal logical inferences, it is "sweepingly convincing" [Lakatos, 1978, p. 64]. By historical examples, Lakatos shows that attempted proofs of a mathematical conjecture are first accepted, then criticized, rejected, transformed, improved and newly criticized. But this process of mathematical growth would not take place if the counterexamples given to criticize attempted proofs were not perceived as compelling. Lakatos denies that real mathematical proofs are indestructible chains of deductive links yielding absolute certainties, but he does not abandon the idea that proofs or, at least, fragments of proofs, can be compelling.

2 Is there such thing as a substantially compelling inference?

The preceding remarks suffice for justifying my topic. A philosophical study of proofs should deal with the compulsion of proofs. An important feature of the traditional notion of compulsion is that an argument can exercise its compelling force "in opposition to everything that is commonly said" [Plato, 1997, 361], against the convictions and the inclinations of the members of the community, so that also the community, and not only individual subjects, can be compelled. Though very reluctant, the community of Pythagoreans was forced to accept the proof that there are incommensurable magnitudes. I call the compulsion which constrains a community "substantial compulsion". How is substantial compulsion possible? Is it really possible? On today's philosophical agenda one of the problems is whether the compulsion of proofs is only a myth. The problem is raised by Ludwig Wittgenstein in his *Remarks on the Foundations of Mathematics*.

> "But am I not compelled, then, to go the way I do in a chain of inferences?" — Compelled? After all I can presumably go as I choose! [Wittgenstein, 1956, I, §113]

In Wittgenstein's writings a line of thought has been found which seems to show that the notion of a substantial compulsion of proofs is empty. The first to describe this line of thought was Michael Dummett [1959b, cf. 1990]. In the sequel I shall refer to it as "the Wittgensteinian critique". My concern, however, is not to argue for an interpretation of Wittgenstein and to refute other interpretations. The critique is "Wittgensteinian" at least in the sense that it is inspired by Wittgenstein's writings. To illustrate the Wittgensteinian critique, I shall first attempt an analysis of the notion of compulsion.

3 Normative compulsion: good inferences.

Since the compulsion of a proof depends on whether the component inferences are compelling, let me facilitate my task by trying to clarify the notion of a compelling *inference*. Consider the following argumentation steps:

(i) the area of square D is the double of the area of square A
∴ a side of D is the double of a side of A;

(ii) square B can be divided into four squares, each of which is equal to A
∴ B is four times A.

Both inferences are treated by the boy in the *Meno* as immediately obvious. But the exchange with Socrates reveals the difference. Inference (i) is wrong and leads the boy to a false conclusion. Inference (ii) is a compelling inference that forces the boy to reject (i). Subjectively, the inclination the boy initially feels for the first inference may be as strong as the inclination felt for the second inference. Nevertheless, if we adopt the traditional notion of compulsion, illustrated by Plato, we should not put the two inferences on the same level. In performing (i) the boy may perhaps feel compelled, but he is *not* really compelled, because the inference is wrong: if the boy were not corrected by Socrates, (i) would avert him from the truth. Inference (ii), instead, brings him nearer to the truth on the investigated matter. For a subject to be compelled by an inference, it is not only necessary that the subject believes that the inference brings him, or her, nearer to knowing the truth. It is also necessary that the inference really has the property of being truth-conducive. This does not mean that the inference has to be deductive, i.e. necessarily truth-preserving. An inference may be compelling even if it is not deductive. My point here is not that a compelling inference has to be deductive, but that there is a normative ingredient in our inferential activity and the normative ingredient is constitutive of the notion of compulsion too. Drawing inferences is not a pointless game. Even though our personal aims in performing inferences can be of many kinds, it is part of the nature of inferential practice that we consider inferential acts responsible for their being good means of approaching truth. If they are not, we think they deserve to be criticized: they are bad inferences. So we can distinguish good and bad inferences. *A "good inference" is an inference that we ought to accept if we are interested in truth.* Only good inferences can be really compelling.

4 Natural compulsion?

The normative ingredient poses an obstacle for a naturalistic or causal explanation of the notion of compulsion. Psychologists have variously established that human subjects in certain contexts have the natural tendency to perform bad inferences. Evidence to this effect is given, for instance, by Wason's experiments testing the ability to falsify conditional hypotheses (cf. [Wason, 1966]). These experiments show that if we defined a compelling inference as an *inference that subjects are naturally inclined to perform or accept*, many easily recognizable fallacies would have to be considered compelling. A plausible naturalistic characterization of a compelling inference should thus be more sophisticated. Kenneth Taylor suggests a naturalistic explanation of "the compulsion of reason" in terms of norms that a "cognizer" would endorse "upon competent reflection" [Taylor, 2000, p. 235]. Taylor says "competent reflection" and "endorsement" are "purely psychological or functional role concepts" [Taylor, 2000, p. 236]. This explanation of compulsion, however, cannot be adequate if a competent reflection is not meant as a reflection *aiming at truth*. Taylor has only shifted the problem: we would have a naturalistic explanation of inferential compulsion if we had an acceptable naturalistic theory of "competent reflection aiming at truth".

5 Minimal objectivity.

The normativity of inferential activity reveals itself in our acknowledging a distinction between good and bad inferences. We want to account for the intuition that sometimes, though we are inclined to accept an inference, the inference is bad. This intuition would turn out to be misleading, if we in the end came to endorse the view that inferences are made good by their being merely treated as good. To treat an inference as a good inference is to be inclined to accept it in contexts where one would describe oneself as aiming at truth. If we take the normativity of inferential activity seriously, we should think that *being treated* as a good inference and being a good inference are not the same property. In other words, we should endorse *the minimal objectivity of good inference*, which can be so stated:

(a) argumentation steps can be treated as good inferences without being good inferences and

(b) argumentation steps can be good inferences without being treated as good inferences.

6 Argumentation steps.

To clarify the property of being a compelling inference something must be said about the bearers of this property. Hence some words are in order concerning argumentation steps. My notion of "argumentation step" is a generalization of Prawitz's notion of "inference" (cf. [Prawitz, 1973, p. 228]). It is important to understand that an argumentation step is not simply a symbolic configuration, but an act in which a symbolic configuration is used. An argumentation step is the particular act of giving evidence for a token sentence, called conclusion (possibly depending on certain hypotheses). At least seven features *can* be relevant to describe an argumentation step I: the conclusion C, some non-linguistic evidence (e.g. a drawing), linguistic premises, arguments for those premises, assumptions discharged by I, free variables which are bound by I and, finally, an indication of whether I is meant as a conclusive or a defeasible step. Only the latter indication and an indication of the conclusion always identify relevant features of the argumentation step. The other ingredients can be absent (e.g. when axioms are put forward) (cf. [Cozzo, 1994, pp. 59–63]). If the non-linguistic evidence is absent, the argumentation step can be called a *pure* inference. However, the word "inference" will be henceforth used in a broad sense, as equivalent to "argumentation step". The detailed characterization of argumentation steps is not pedantic in the present context, because the same symbolic configuration can constitute premises and conclusion of two argumentation steps that are different with respect to some of the other aforementioned features, so that one of the two steps is a good inference while the other is a bad inference. For example, the same premises and conclusion can be employed in a good defeasible argumentation step, or in a bad deductive step meant as conclusive.

7 Minimal compulsion.

Taking the normativity of inferential activity into account, we can define compulsion as a relation between an argumentation step I and a speaker-reasoner S. The first clause is about the inclination of S; the second about the goodness of I; the third about the right connection between goodness and inclination.

An argumentation step I is *compelling* for S if, and only if,

1. S is, or would be, (predominantly) inclined to accept I, if S is, or were, (predominantly) interested in truth;

2. I is a good inference (therefore S ought to accept I, if S is, or were, interested in truth);

3. S (implicitly) knows that I is a good inference, and in suitable conditions would give (2) as an explanation for the inclination in (1).

Clause (2) has to be added, and does not follow from clause (1), nor does clause (1) follow from (2), if and only if the minimal objectivity of good inferences holds. If it holds, we have to add also clause (3), because S's inclination and the goodness of I are independent of each other, and it is possible that the motive of S's inclination is not his or her knowing that I is good. In certain circumstances S can be inclined to accept I for a wrong reason (e.g. someone's authority or other psychological influences lead S to accept I) and in such circumstances, even if I is a good inference, we shall not say that S is compelled by I. A *minimal* notion of compulsion has instances only if the following *thesis of compulsion* is true:

> for some subject S and some argumentation step I, I is compelling for S, and the minimal objectivity of good inference holds.

We can ask whether it ever happens that S is compelled by an argumentation step I in this sense. It does not seem difficult to find cases in which clause (1) is satisfied. We often sincerely describe ourselves as wanting to know the truth about an investigated matter and we are ready to say that we accept a particular inference because this is what one should do if one is interested in coming nearer to the truth. But, if the objectivity of good inference holds, such self-descriptions might be self-deceptions, because the argumentation step I might be an inference which is not a good inference (thus (2) would be false) or because we do not know that I is a good inference (in this case (3) would be false). Therefore, if the objectivity of good inference holds, we have to face the task of explaining how one can know that an inference is a good inference. There is, however, a radical objection against the thesis of compulsion: one can deny both conjuncts (a) and (b) of the objectivity of good inference. If the objectivity of good inference is abandoned, the thesis of compulsion is false. We cannot anymore distinguish between an inference which is compelling for S and an inference which S is only inclined to accept. The whole idea of a compulsion of proof would reduce to the fact that we sometimes *feel* compelled or *believe* ourselves to be compelled.

8 Social compulsion.

The Wittgensteinian Critique does not deny compulsion, but reduces it to a social compulsion. In *Remarks on the Foundations of Mathematics* we find the following example of "the special activity of inferring":

> A regulation says "All who are taller than five foot six are to join the ... section". A clerk reads out the men's names and heights. Another allots them to such-and-such sections. "N.N.

five foot nine." "So N.N. to the ... section." That is inference. [Wittgenstein, 1956, I, §17].

The abovementioned is a logical inference, but for Wittgenstein the "inexorability of mathematics" is "the same as that of logical inference" [Wittgenstein, 1956, I, §§4–5]. The example is used to illustrate the view that compulsion is social compulsion:

> the laws of inference can be said to compel us; in the same sense, that is to say, as other laws in human society. The clerk who infers as in (17) *must* do it like that; he would be punished if he inferred differently. If you draw different conclusions you do indeed get into conflict, e. g. with society; and also with other practical consequences [Wittgenstein, 1956, I, §116].

> In what sense is logical argument a compulsion? — "After all you grant *this* and *this*; so you must also grant *this*!" That is the way of compelling someone. That is to say, one can in fact compel people to admit something in this way. Just as one can e. g. compel someone to go over there by pointing over there with a bidding gesture of the hand. [Wittgenstein, 1956, I, §117]

The Wittgensteinian idea is not that the compulsion of inferences is based upon the explicit social stipulation and overt enforcement of certain rules, but that the only basis for a distinction between good and bad inference is a social agreement, which "is not an agreement in opinions or convictions" [Wittgenstein, 1956, VI, §30], "but in the form of life" [Wittgenstein, 1953, I, §241]: all, or almost all, members of the relevant community simply *act* in this way. A deviating individual, who does not accept an inference which is accepted by the community, gets into conflict with society or is excluded from it.

> This is a demonstration for whoever acknowledges it as a demonstration. If anyone doesn't acknowledge it, doesn't go by it as a demonstration, then he has parted company with us even before anything is said [Wittgenstein, 1956, I, §61].

The Wittgensteinian line of thought adopts a *weak version* of the objectivity of good inference, which binds only single individuals:

(a) argumentation steps can be treated as good inferences by individual subjects without being good inferences and

(b) argumentation steps can be good inferences without being treated as good inferences by individual subjects.

9 The relativity of good inference.

We experience compulsion most significantly when we have already accepted a piece of reasoning, but a subsequent compelling argument forces us to admit that our previous reasoning was wrong. On the face of it, such a revision of our inferential attitudes embodies the idea that being a good inference does not follow from being accepted by us. To the extent that goodness is independent of acceptance, the requirement that our inferences be good, imposes an independent constraint upon our inferential practice. However, it seems that, according to the social conception of compulsion, while an individual's inferential act can be wrong, the community's acceptance of an inference is *ipso facto* right, because it *constitutes* rightness. This is true in so far as the social conception of compulsion involves the thesis of *relativity of good inference*:

> an argumentation step is a good inference if, and only if, it is treated as a good inference by the community.

If good inference is relative to a community, and being a good inference is the same as being accepted by the community, no genuine normativity holds for the community: the community can never be wrong. We cannot distinguish between the community's feeling compelled and its being compelled. This is why it seems fair to say that the notion of a compulsion which constrains the community dissolves. As Michael Dummett writes:

> to adopt this picture involves thinking with Wittgenstein that we are *free* in mathematics at every point; no step we take has been forced on us by a necessity external to us, but has been freely chosen [Dummett, 1959a, p. 18].

Dummett's point here is not that, according to the Wittgensteinian, the *experience* of performing or accepting an inference is the same as that of arbitrarily choosing one of several possible courses of action. Nor does Dummett mean that the Wittgensteinian regards an individual's acceptance of an inference as the result of a sort of social investigation about what fellow speakers have chosen to accept. The members of the community *feel* compelled *by the inference*, and would not describe their act as "deciding that the inference is correct". But there is nothing beyond the "brute fact" [Dummett, 1990, p. 449] that they feel compelled, nothing that makes the inference a good inference independently of the community's acceptance. One cannot *explain* our compulsion by saying that we feel compelled because we know the independent fact that the inference is valid. Nothing justifies our feeling of compulsion associated with an inference. If members of an alternative community (in all other respects similar to us) had a different feeling and rejected the inference, they would "part company" with us; we would say that they are wrong, but nothing would make their attitude wrong independently of our judgement and feeling, which count only for us. For them rejecting the inference would be right. Friedrich Nietzsche would have commented:

> One is always wrong, but with two, truth begins. One cannot prove his case, but two are irrefutable [Nietzsche, 1974, p. 260].

10   Community-independent objectivity.

The relativity of good inference is counterintuitive. Counterintuitive, in particular, is the view that a community can never be wrong. It often happens that a criticism is addressed against a proof previously accepted by the mathematical community. Sometimes the criticism succeeds. The result is that the proof is rejected. Alfred Bray Kempe offered a proof of Francis Guthrie's four colours conjecture in 1879. Eleven years later the community agreed to the criticism of Percy John Heawood and Kempe's argument was not considered valid anymore. If the relativity of proof holds, similar events are *only* changes of the community's attitude; they should not be described as the discovery of an objective mistake. In order to avoid such counterintuitive consequences, we have to deny the relativity of good inference, which amounts to adopting a *community-independent objectivity of good inference*:

  a) argumentation steps can be treated as good inferences by the community without being good inferences and

  b) argumentation steps can be good inferences without being treated as good inferences by the community.

The *thesis of substantial compulsion*, according to which also a community can be compelled by a constraint which is independent of the community's actual acceptance, is the following:

> for some subject S and some argumentation step I, I is compelling for S, and the community-independent objectivity of good inference holds.

The Wittgensteinian critique seems to undermine the community-independent objectivity of good inference and hence the thesis of substantial compulsion. Wittgenstein's remarks threaten community-independent objectivity, in that they launch an attack on a widespread conception of meaning.

11 The complete determination of meaning.

How can the community be wrong? What can make an inference good, though the community rejects it? What can make an inference bad, though the community accepts it? What kind of constraint can limit the community's freedom? Dummett's answer is: the meanings of the involved expressions. Premises and conclusions of our inferences have a determinate meaning. They concern the areas of squares, or map colouring, or natural numbers. The community has fixed the meanings. But the statements in question do not deal with the community's attitudes. They deal with squares, planar graphs, numbers. Following Dummett, one can maintain that "we have no alternative" but to consider certain inferences good inferences, "if we are to remain faithful to the understanding we already had of the expressions contained in [them]" [Dummett, 1959b, p. 173]. The reason is that the meanings of the involved expressions determine inferential goodness in advance, independently of the way we shall actually treat inferences when we encounter them. To inferential uses of language we are applying a more general view, the *thesis of complete determination of meaning*:

> after a finite process of meaning-fixation relative to an expression X, the meaning of X is completely fixed, in such a way that the correctness (or incorrectness) of future uses of X in new unconsidered circumstances is determined *in advance*.

According to the complete determination thesis, future uses of an expression X are correct only if they are faithful to the already fixed meaning of X, and whether they are faithful or not is determined in advance, when the meaning is attached to X. Before considering the statements "B is four times A" and "square B can be divided into four squares, each of which is equal to A", the English speaking counterparts of Socrates and the slave, have attached to the involved words certain meanings. They have conferred a certain meaning to "four" and "times", have associated A and B with two different geometrical constructions, and so on. By knowing these meanings, the speakers-reasoners understand the statements. But the meanings are not only constitutive of the speakers' understanding. They also determine the truth-conditions of the two statements and the correctness of inferring "B is four times A", from "square B can be divided into four squares, each of which is equal to A" (inference (ii) in § 3). The involved meanings determine the correctness of this inference in advance, before it is performed. Though a compelling inference is not necessarily deductive, elementary deductive inferences are a typical example of compelling inferences. If the thesis of complete determination of meaning holds, the involved meanings make (ii) a deductively valid inference, before we encounter (ii) and judge whether it is valid; hence (ii) preserves truth, independently of our endorsement. This is why, if we are interested in truth, we ought to conform our judgement to the already existing conceptual connection between the premise and the conclusion and thus accept (ii). The idea that there is an already existing connection that imposes itself upon us was also part of Plato's view that the slave was recollecting a geometrical knowledge already possessed before birth. Recollection apart, Wittgenstein describes a similar view in the following remark about Russell and Frege:

> Russell seems to be saying of a proposition: "It already follows — all I have to do is, to infer it". Thus Frege somewhere says that the straight line which connects any two points is really already there before we draw it; and it is the same when we say that the transitions, say in the series + 2, have really already been made before we make them orally or in writing — as it were tracing them [*gleichsam nachzögen*]. [Wittgenstein, 1956, I, §21].

Like archaeologists tracing the course of an ancient road, we may be misled and go the wrong way. We may reject a good inference. Since the inference is *already* good, we are wrong. Hence there is a constraint that limits our freedom. It is we who fix the meanings of words, but once they are fixed, meanings impose upon us standards of correctness which are independent of our judgement. Assuming the complete determination thesis, we can explain the fact that an inferential step I is compelling for a subject S independently of the community's attitude. If the thesis is true, meaning determines goodness. Thus, we can explain why, given the meanings of the statements involved in I, the latter is a good inference (clause (2) of the definition of compulsion in §7). On the other hand, if S understands those statements, S knows their meanings. Knowledge of meaning, if meaning determines goodness, may account for the fulfilment of clause (3), i.e. for the fact that S knows that I is a good inference. And the latter knowledge may explain the fulfilment of clause (1), i.e. the fact that S is, or would be, inclined to accept I, if S is, or were, interested in truth.

12 The Wittgensteinian Critique

The complete determination thesis expresses a conception of meaning according to which meanings implicitly involve rules or norms fixing the correctness of linguistic uses in advance. This conception of meaning implies the community-independent objectivity of good inference and the thesis of substantial compulsion: it constitutes a theoretical frame in which substantial compulsion can be explained and becomes intelligible. The force of the Wittgensteinian critique of substantial compulsion depends on its attack on the complete determination thesis, which destroys this hospitable frame. Complete determination can be viewed as an application to meanings of a general conception of rules, which is the target of Wittgenstein's rule following considerations:

> "Once you have got hold of the rule, you have the route traced for you" [Wittgenstein, 1956, VI, §31].

> we might imagine rails instead of a rule. And infinitely long rails correspond to the unlimited application of a rule [Wittgenstein, 1953, I, §218].

> I no longer have any choice. The rule, once stamped with a particular meaning, traces the lines along which it is to be followed through the whole of space. [Wittgenstein, 1953, I, §219].

The supposed fact that we understand the meaning of a word in this sense is what Wittgenstein calls *eine übermässige Tatsache*, a superlative fact that "in a flash" contains "the whole use of the word" [Wittgenstein, 1953, I, §191]. How can a fact of any kind determine the correctness of infinitely many potential uses of a word in advance? What can constitute a fact endowed with the non-causal (and non-contingent) power of predetermining correctness or incorrectness of unconsidered future uses? If we try to explain what a fact of this kind might consist in, our attempts end in failure (cf. [Kripke, 1982]). We cannot give any non-circular specification of features of our behaviour or mental life that could constitute such a fact. We thus face a dilemma. Either we maintain that, after all, the notion of such a fact is fully intelligible and does not require any explanation, or we agree with Wittgenstein, that we "have no model of this superlative fact" [Wittgenstein, 1953, I, §192] and that the corresponding conception of meanings and rules is only a "mythological description" [Wittgenstein, 1953, I, §221] which should not be taken seriously as a philosophical explanation. The first attitude seems to me dogmatic. Therefore I prefer the second alternative. The second alternative leads to rejecting the thesis of complete determination. Does this mean we have to reject substantial compulsion too?

13 Plasticity of meaning.

The sceptic described by Saul Kripke in *Wittgenstein on Rules and Private Language* would answer in the affirmative. From the rule following considerations Kripke's sceptic concludes that "there can be no such thing as meaning" [Kripke, 1982, p. 54, cf. p. 71]. If there is no fact as to whether we mean something by premises and conclusions of our inferences, there seems to be nothing from which substantial compulsion can originate. However, rejecting complete determination does not necessarily involve a rejection of the notion of meaning. We may adopt a different conception of meaning, which *I* propose to call *"the plasticity of meaning"*:

> the meaning of X is never completely determined or fixed: it is continuously moulded and shaped by common use (i.e. use accepted by the community).

The plasticity of meaning agrees with Wittgenstein's idea that concepts are "pliable" [Wittgenstein, 1956, IV, §4]. Crispin Wright seems to have this view in mind when he highlights "the capacity of ongoing use to determine meaning" [Wright, 1986, p. 274]. If we adopt the plasticity of meaning, the problem arises whether there is any remaining sense in which we can say that substantial compulsion and community-independent objectivity hold.

14   From plasticity to relativity.

The problem arises because an argument can be given from the plasticity of meaning to the relativity of good inference.

(i) The meaning of X is never completely determined or fixed, it is continuously moulded and shaped by common use.

(ii) Every common use of X is constitutive of the meaning of X.

(iii) Every common use of X is correct.

(iv) If the community treats (or does not treat) an argumentation step involving X as a good inference, the community is right.

(v) Argumentation steps are good inferences if, and only if, they are treated as good inferences by the community.

If this reasoning is right, substantial compulsion falls as in a domino effect. The complete determination of meaning falls. But community-independent objectivity depends on complete determination. So, community-independent objectivity falls too. Hence substantial compulsion falls. The outcome of the Wittgensteinian critique is that we can make sense of a social compulsion according to which a single individual can be wrong, but we cannot make sense of the community's being wrong, and thus we cannot make sense of the community's being substantially compelled. However, I think reasoning (i)–(v) is wrong. There are two mistakes.

15   Primitive plasticity

A first mistake is the inference from (i) to (ii). Plasticity, formulated in (i), does not say that the meaning of X is moulded and shaped by *all* common use. Hence it does not follow that every common use of X is constitutive of the meaning of X. The property of being constitutive of meaning manifests itself in the way certain uses are treated. This allows us to characterize *primitive* uses.

> A use U of X in a language L is a *primitive* use, if, and only if, speakers of L treat any deviation with respect to U as indicating a lack of understanding of X.

For example, an inference from "A is a square" to "A has four sides" is treated as primitive. We normally act as if accepting this inference were simply part of what counts as understanding "square" and therefore the inference did neither admit nor require any justification. On the other hand, the inference from "the area of square D is the double of the area of square A" to "a side of D is equal in length to the diagonal of A" is not a primitive use. Thus we can formulate a more precise version of plasticity, the thesis of *primitive plasticity of meaning*:

> the meaning of X is never completely determined or fixed: it is continuously moulded and shaped by common primitive uses of X (and possibly of some other related words). Non-primitive uses are not constitutive of meaning.

16 Plasticity and substantial compulsion

Primitive plasticity of meaning is compatible with the objectivity of good inference. To substantiate my claim, I describe two possible epistemic courses. The first epistemic course consists of five stages. First stage: the community treats an inference I^1 as a good inference. Second stage: a dissident d is sceptical about I^1 and calls on those who accept I^1 to further elaborate and articulate the step from premises to conclusion. Third stage: attempts at elaborating and articulating I^1 show that something is wrong, e.g. by drawing attention to some clash between I^1 and other uses. Fourth stage: the community admits that d's doubts were right and rejects I^1. Fifth stage: the community admits that I^1 is not and never was a good inference. Thus, it seems fair to conclude that, at the first stage of this epistemic course, I^1 is treated as a good inference by the community, but it is not a good inference. Also the second epistemic course consists of five stages. First stage: a single mathematician m presents an inference I^2 as a good inference. Second stage: the community, after examination of I^2, does not accept I^2, which is put aside. Third stage: another mathematician m^* reproposes I^2: he (or she) highlights the structure of I^2, articulates it and shows that it can be justified in terms of other uses. Fourth stage: the community gradually comes to agree that m was right. Fifth stage: the community accepts I^2 as a good inference and admits that it was a good inference also at the second stage, though its being a good inference was not understood yet. Thus it seems fair to conclude that, at the second stage of this epistemic course, I^2 is a good inference even if I^2 is not treated as a good inference by the community. Both stories are not uncommon in mathematical practice. They illustrate that the community implicitly acknowledges a principle that might be called *"principle of self-corrigibility"*:

> if, under pressure of critical discussion, we articulated our judgments about some uses, we could discover that these judgments are wrong and then we ought to correct them.

That the community acknowledges self-corrigibility does not in itself show that we can make sense of the idea that the community is wrong. The community's endorsement of self-corrigibility seems paradoxical if one espouses the view that good inference and truth are constituted by the community's acceptance. But the air of paradox surrounding that view is not enough to substantiate the objectivity of good inference. The paradoxicality of relativism does not provide us with a picture of our practice that makes the possibility that the community is wrong intelligible. To make it intelligible we have to show that an adequate conception of meaning is available, into which the community-independent objectivity of good inference can be embedded without contradiction or implausibility. Is primitive plasticity such a conception of meaning? The conclusions of the two stories are precisely the two conjuncts of the thesis of community-independent objectivity. Are they compatible with the primitive plasticity of meaning? They certainly are if I^1 and I^2 are non-primitive uses: critical investigation can clearly lead to rational changes in our attitude towards argumentation steps which are not constitutive of meaning. What if I^1 and I^2 are primitive uses? The latter question leads me to the second mistake in reasoning (i)–(v) of §14: the inference from (ii) to (iii). The inference depends on the hidden assumption that if a use contributes to the meaning of an expression, then it must be correct. This erroneous assumption is a version of the doctrine of analytical validity as validity in virtue of meaning. It appears plausible if one thinks that the activity of meaning-shaping is completely free and immune to criticism. If this were the case, meaning-constitutive uses would be beyond rational critique. However, it is not the case. Suppose I^1 is treated as a primitive use by the community. In the first epistemic course the dissident calls on interlocutors to further elaborate and articulate I^1, and thereby throws a meaning-constitutive principle into question. It can very well happen that the resulting critical investigation reveals that something is wrong. An example is the assertion of the principle of unrestricted comprehension: "one can always collect together into a set all the things satisfying a given description", which was a primitive use constitutive of the meaning of "set" in Cantor's theory of sets. The discovery of set-theoretical antinomies led, in the end, to rejection of this primitive use. To criticize a primitive use is to criticize a fragment of language and the involved concepts. We know from the history of science that this is sometimes a decisive move. As far as mathematics is concerned, in his book *Proofs and Refutations*, Lakatos has called "concept-stretching" [Lakatos, 1976, pp. 83–99] an instance of such a criticism and the modifications and improvements of the language that arise from it. In the course of mathematical rational controversy accepted meanings (i.e. concepts) can be criticized and new meanings (i.e. concepts) can be put forward. In the same book Lakatos has vividly illustrated that new concepts (i.e new meanings) can be generated by critical mathematical investigation. In the second epistemic course m may present a new concept with respect to which the inference I^2 is a primitive use, and the new concept can turn out to be so fruitful that I^2 is finally accepted by the community, even though it was at first rejected. For Michael Crowe it is a law concerning change in the history of mathematics that "many new mathematical concepts [...] meet forceful resistance after their appearance and achieve acceptance only after an extended period of time" ([Crowe, 1975], in [Gillies, 1992, p. 16]). An example is the response of the mathematical community to the proposal of square roots of negative quantities. Crowe adds the law that "although the demands of logic, consistency, and rigour have at times urged the rejection of some concepts now accepted, the usefulness of these concepts has repeatedly forced mathematicians to accept and to tolerate them, even in the face of strong feelings of discomfort" (p. 17). The word "forced" suggests that compulsion is a feature of these mathematical changes. A confirmation comes from Crowe's quotation of Felix Klein: "imaginary numbers made their own way [...] without the approval, and even against the desires of, individual mathematicians, and obtained wider circulation only gradually and to the extent to which they showed themselves useful" [Klein, 1939, p. 56]. The concept of "imaginary number" and the primitive uses of corresponding expressions were rejected at first and then gradually accepted, because of their epistemic fruitfulness. If we consider the acceptance of imaginary numbers a rational change, as I think we should, we can conclude that the second epistemic course can give rise to a rational change even if I^2 is a primitive use. A meaning-constitutive use is not beyond rational criticism and investigation. Critical investigation can lead to meaning-changes and can generate new meanings. The conclusions of the two stories, and hence the thesis of community-independent objectivity of good inference, are compatible with the primitive plasticity of meaning independently of whether the inferences in question are primitive or non-primitive uses. We have thus established that substantial compulsion can hold even if meaning is plastic. One might object that our notion of compulsion is a problematic hybrid of two independent ingredients, the inclination of a subject S to accept the argumentation step I, and the fact that I is a good inference. The objection does not take into account the third necessary condition of compulsion that connects the two ingredients: the compelled subject must *know* that I is a good inference. The goodness of I is characterized here in terms of its being the result of the epistemic process of critical investigation aiming at truth. A good inference is an inference that brings us nearer to knowing the truth on the relevant matter. I am assuming that critical investigation does bring us nearer to truth and that through critical investigation we can know that an inference is good. Such an assumption is fully justified, if we endorse an epistemic conception of truth. According to the epistemic conception of truth I have in mind, a true statement is precisely a statement that critical investigation, if it were carried far enough, would lead us to accept in such a way that no further information would bring about a rational revision concerning that statement. If truth is epistemic in this sense, also the notion of good inference is epistemic: a good inference is an inference to which we can be led by critical investigation. Critical investigation can bridge the gap between inclination and goodness. Goodness is independent of inclination, but epistemically accessible.

17 Conclusions

The present paper offered reasons to deny that the Wittgensteinian critique starting from the rule-following considerations really undermines the idea of a compulsion of proofs. The paper describes a plausible view in which the plasticity of meaning and the community-independent objectivity of good inference cohabit without contradiction. The claim that the community can be wrong, and compelled by a proof to change shared convictions, does not require endorsement of the mythological conception of a completely fixed meaning which predetermines all the correct uses of an expression. Even if we think that meaning is continuously shaped by our common primitive uses, we can reasonably maintain that the community is not free in mathematics. Inferential steps can be forced on us, but they are not forced by a necessity external to us: they are forced by our common critical activity.

18 Acknowledgements

For his valuable comments on an earlier version of this paper I would like to thank Professor Dag Prawitz. I am also grateful to Professor Carlo Cellucci for stimulating discussions on the problems here dealt with.

BIBLIOGRAPHY

[Cozzo, 1994] Cozzo, C. *Meaning and Argument.* Stockholm: Almqvist & Wiksell, 1994.
[Crowe, 1975] Crowe, M. Ten 'Laws' concerning patterns of change in the history of mathematics, *Historia Mathematica,* **2**, 161–166, 1975. Rep. in [Gillies, 1992, pp. 15–20].
[Dummett, 1959a] Dummett, M. Truth. In [Dummett, 1978, pp. 1–19].
[Dummett, 1959b] Dummett, M. Wittgenstein's philosophy of mathematics. In [Dummett, 1978, pp. 166–185].
[Dummett, 1978] Dummett, M. *Truth and Other Enigmas,* London: Duckworth, 1978.
[Dummett, 1990] Dummett, M. Wittgenstein on necessity: some reflections. In [Dummett, 1993, pp. 446–461].
[Dummett, 1993] Dummett, M. *The Seas of Language,* Oxford: Clarendon Press, 1993.
[Gillies, 1992] Gillies, D., ed. *Revolutions in Mathematics,* Oxford: Oxford University Press, 1992.
[Klein, 1939] Klein, F. *Elementary Mathematics from an Advanced Standpoint: Arithmetic, Algebra, Analysis,* Dover, New York, 1939.
[Kripke, 1982] Kripke, S. *Wittgenstein on Rules and Private Language,* Harvard University Press, Cambridge, Massachusetts, 1982.
[Lakatos, 1976] Lakatos, I. *Proofs and Refutations,* Cambridge University Press, Cambridge, 1976.
[Lakatos, 1978] Lakatos, I. 1978, *Mathematics, Science and Epistemology, Philosophical Papers,* Vol. 2, ed. by J. Worrall and G. Currie, Cambridge University Press, Cambridge.
[Nietzsche, 1974] Nietzsche, F. *The Gay Science,* ed. by W. Kaufmann, Vintage, New York, 1974.
[Plato, GO] Plato. *Gorgias.* In [Plato, 1997, pp. 791–869].
[Plato, ME] Plato. *Meno.* In [Plato, 1997, pp. 870–897].
[Plato, PA] Plato. *Parmenides.* In [Plato, 1997, pp. 359–397].
[Plato, 1997] Plato. *Complete Works,* ed. by J. M. Cooper. Hackett, Indianapolis/Cambridge, 1997.
[Prawitz, 1973] Prawitz, D. Towards a foundation of a general proof theory. In [Suppes *et al,* 1973, pp. 225–250].
[Suppes *et al,* 1973] Suppes, P. *et al.,* (eds). *Logic, Methodology and Philosophy of Science,* Volume IV. North Holland, Amsterdam, 1973.
[Taylor, 2000] Taylor, K. A. What in nature is the compulsion of reason? *Synthese,* **122**, 209–224, 2002.
[Wason, 1966] Wason, P. Reasoning. In B. M. Foss (ed.), *New Horizons in Psychology,* pp. 135–151. Penguin, Harmondsworth, 1966.

[Wittgenstein, 1953] Wittgenstein, L. *Philosophical Investigations*, Blackwell, Oxford, 1953.
[Wittgenstein, 1956] Wittgenstein, L.: 1956, *Remarks on the Foundations of Mathematics*, ed. by G.H. von Wright, R. Rhees, G.E.M. Anscombe, Blackwell, Oxford.
[Wright, 1986] Wright, C. Rule following, meaning and constructivism. In C. Travis (ed.) *Meaning and Interpretation*, pp. 271–297, Blackwell, Oxford, 1986. Rep. in [Wright, 2001].
[Wright, 2001] Wright, C. *Rails to Infinity*, Harvard University Press, Cambridge, 2001.

www.ingramcontent.com/pod-product-compliance
Ingram Content Group UK Ltd.
Pitfield, Milton Keynes, MK11 3LW, UK
UKHW041430180426
11947UKWH00007B/375